Anna Näslund, Ramón Reichert (eds.)
Digital Culture & Society (DCS)

Digital Culture & Society | Volume 18

Editorial

Digital Culture & Society is a refereed, international journal, fostering discussion about the ways in which digital technologies, platforms and applications reconfigure daily lives and practices. It offers a forum for critical analysis and inquiries into digital media theory. The journal provides a publication environment for interdisciplinary research approaches, contemporary theory developments and methodological innovation in digital media studies. It invites reflection on how culture unfolds through the use of digital technology, and how it conversely influences the development of digital technology itself.
The journal is edited by Mathias Fuchs, Ramón Reichert and Anna Näslund.

Board:
Maria Bakardjeva (University of Calgary), David Berry (University of Sussex), Jean Burgess (Queensland University of Technology, Brisbane, Australia), Mark Coté (King's College London), Colin Cremin (University of Auckland), Sean Cubitt (Goldsmiths, University of London), Mark Deuze (University of Amsterdam), José van Dijck (Utrecht University), Delia Dumitrica (Erasmus University Rotterdam), Astrid Ensslin (University of Alberta, Edmonton), Sonia Fizek (Abertay University), Federica Frabetti (University of Oxford), Orit Halpern (The New School, New York), Irina Kaldrack (Braunschweig University of Art / Leuphana University of Lüneburg), Denisa Kera (National University of Singapore), Lev Manovich (The Graduate Center, The City University of New York), Janet H. Murray (Georgia Institute of Technology, Atlanta), Jussi Parikka (University of Southampton), Lisa Parks (University of California, Santa Barbara), Dominic Pettman (The New School, New York), Rita Raley (University of California, Santa Barbara), Richard Rogers (University of Amsterdam), Julian Rohrhuber (Robert Schumann School of Music and Media, Düsseldorf), Marie-Laure Ryan (University of Colorado, Boulder), Mirko Tobias Schäfer (Utrecht University), Jens Schröter (University of Bonn), Trebor Scholz (The New School, New York), Tamar Sharon (Maastricht University), Roberto Simanowski (City University of Hongkong), Nathaniel Takcz (University of Warwick), Geoffrey Winthrop-Young (University of British Columbia, Vancouver), Sally Wyatt (Maastricht University)

Anna Näslund (former Dahlgren) is professor of art history at Stockholm University. She has written extensively on different aspects of visual culture, photography, media history, digitization, archives and museum practices. She has been the PI of several research projects, such as »The Politics of Metadata« and »Sharing the Visual Heritage« focusing different aspects of cultural heritage institutions image collections online, and media historical projects on social media photography and photo albums.
Ramón Reichert (Dr. phil. habil.) teaches and researches as a senior researcher at the Department of Cultural Studies at the Universität für Angewandte Kunst in Vienna. Previously, he taught and researched in Basel, Berlin, Canberra, Fribourg, Helsinki, Sankt Gallen, Stockholm and Zurich and was EU project coordinator for many years. His current research project »Visual Politics and Protest. Artistic Research Project on the visual framing of the Russia-Ukraine War on internet portals and social media« (2022-2024) deals with the visual politics of violence, conflict and resistance.

Anna Näslund, Ramón Reichert (eds.)

Digital Culture & Society (DCS)

Vol. 10, Issue 1/2024 – Digital War: Media Strategies and Visual Politics during the Full-Scale Attack of Russia on Ukraine

[transcript]

Funding:

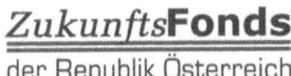
Zukunftsfonds
der Republik
Österreich

Bibliographic information published by the Deutsche Nationalbibliothek
The Deutsche Nationalbibliothek lists this publication in the Deutsche Nationalbibliografie; detailed bibliographic data are available in the Internet at https://dnb.dnb.de

Indexiert in EBSCOhost-Datenbanken.

© 2025 transcript Verlag, Bielefeld
Hermannstraße 26 | D-33602 Bielefeld | live@transcript-verlag.de

All rights reserved. No part of this book may be reprinted or reproduced or utilized in any form or by any electronic, mechanical, or other means, now known or hereafter invented, including photocopying and recording, or in any information storage or retrieval system, without permission in writing from the publisher.

Cover layout: Kordula Röckenhaus, Bielefeld
Typeset: Mark-Sebastian Schneider
Printed by Elanders Waiblingen GmbH, Waiblingen
Print-ISBN 978-3-8376-6868-1 | PDF-ISBN 978-3-8394-6868-5
https://doi.org/10.14361/9783839468685
ISSN of series: 2364-2114 | eISSN of series: 2364-2122

Printed on permanent acid-free text paper.

Content

Digital War
Media Strategies and Visual Politics during the Full-Scale Attack of Russia on Ukraine
Anna Näslund and Ramón Reichert 7

I Digital Transformations of Visual Politics and Collective Memory

War, Authoritarianism & Technology
The History of the Khatyn Massacre and Its Soviet and Post-Soviet Reproductions
Tatsiana Astrouskaya 17

Izolyatsia: A Factory, But a Different One
Production of Images by the Means of Violence
Lesia Kulchynska 41

II Discourse Analysis and Social Media

Sensitive Subjects
Talking about the Residents of the Territories Occupied by Russia in Official Ukrainian Discourse
Yana Lysenko 57

Living Through Narratives
A Psycholinguistic Study of War Stories by Bohdan Lepky and Today's Ukrainians in Print and Digital Media
Serhii Zasiekin 83

Strategic Deception
Russian Trolls' Attempts to Reframe the Annexation of Crimea for English-speaking Audiences on Twitter/X
Maksim Markelov 97

III Digital Warfare

Strategies and Tactics
A Spinozist Approach to Political Agency Online
Max Kramer   121

When the War Goes Viral
Ukrainian and Russian War Memes
Elena Korowin   137

Biographical Notes   161

Digital War
Media Strategies and Visual Politics during the Full-Scale Attack of Russia on Ukraine

Anna Näslund and Ramón Reichert

Since the full-scale Russian attack on Ukraine since February 24, 2022, warfare on social media and online platforms has introduced a new way of mediatizing war. A constant war-related newsfeed on social media and online platforms has emerged. Against this background the war in Ukraine represents a "fractal war – where you choose to subscribe to your own tailored version of warfare in your feed. This makes it the most personalized war in history" (Hoskins/Shchelin 2023: 449). The subscriber system and the follower principle turn the war in Ukraine into the most individualized war in history. In this ever-growing stream of war images, of killed, wounded, and tortured bodies, along with infographics on body counts (the enumeration of the killed members of the opposing war party), and digital mashups (memes, music videos, etc.) are used to dehumanize the enemy.

On the other hand, a digital archive has been created on online-platforms, messenger-apps and social media. It is supported by the Ukrainian civilian population, deploys smartphones to document war crimes and war damage and collects and networks audiovisual material to provide sources of information for subsequent war crimes tribunals. When civilians are used for legal investigations or to provide criminal evidence for war crimes, their personal data itself provides a target for military operations.

In their 2022 edited book *Radical War: Data, Attention and Control in the 21st Century*, Matthew Ford and Andrew Hoskins are considering that the war in Ukraine is the most networked war in history. It is the first war between two states in Europe that is almost entirely mediated by digital technologies. According to Ford and Hoskins, digital war blurs the boundaries between soldiers and the civilian population.

Participatory War

This observation had already been described by William Merrin in his 2019 *Digital War. A Critical Introduction*, where he speaks of a "participatory war" – "a new mode where networked technologies and online public platforms allow anyone within or outside of a conflict zone to participate in informational war, to tell their story, expose events, offer support and contribute towards or expose propaganda" (218).

For Ukrainians, online-platforms, messenger-apps, and social media are acting as digital archives of military violence. Smartphone technology has led to the multiplication of observation orders. A new era of military intelligence, archiving and mutual reconnaissance has emerged: the app communicates the latest military developments, disseminates air raid warnings, documents war crimes, forwards satellite images for possible targets or lists maps of local air raid shelters. The war in Ukraine is the best documented modern military conflict to date, raising several questions and challenges. The era of smartphones has established the ubiquity of eyewitness testimony. In contrast to traditional war reporting, the content can be transmitted unfiltered and in real time. Anyone can feed these images to the public instantaneously, without editorial supervision and discursive contextualization. A non-centralized digital archive has emerged on online platforms and websites, supported by the Ukrainian civilian population.

The practices of "witnessing war" disseminated on the internet are plenty and diverse. They document destruction and violence on the ground, but they also contain testimonies that are collected on websites, messaging services, social network sites and video platforms. The resulting online archives collect photographs, videos and text-based documents and are not only used by those affected to share their experiences, but are also by academics, human rights activists, NGOs, lawyers, and public prosecutors to initiate war crimes trials later. Methods of ethnography, interview research and research of witness literature are being applied to process the various contents; the more eyewitness documents that substantiate a case, the more reliable appear the existing sources.

Media consumption of war-related content

Since the beginning of the war, an endless stream of images of war has emerged in the newsfeed, uploaded for "regarding the pain of others" (Sontag 2004). Other than on Instagram, TikTok or YouTube, Telegram depictions of violence are less censored, injured bodies are shown in pixelated form, recordings of killings at a distance are constantly distributed using combat footage, explicit, non-pixelated depictions of violence of tortured, abused or dead bodies are distributed by way of links. These videos shared on Telegram to the applause of emoticons are effective due to their visual exhibition value and thus inherit the voyeuristic viewing culture of the early "cinema of attractions" (which, according to Tom Gunning, was primarily focused on visual effect and less on storytelling, 1986).

The number of images grows daily and hourly, bringing together injured, imprisoned, mutilated and dead people, perpetrators, and their victims, who are available at any time and any place in the digital theme park of attractions. While Ukrainians are dehumanized as "Nazis" in Russian Telegram channels, opponents of the war are referred to as "orcs" in Ukrainian channels. Orcs are fictional monsters from Tolkien's *The Lord of the Rings*. For example, an author of the Telegram channel "Ukraine Now" writes about a video showing a captured Russian soldier: "Another lying orc" (Ukraine Now, 5.2.2022). "Orcs" are also mentioned in image comments when Russian soldiers are 'punished,' i.e. killed by drone attacks. These drone images are not only documentary images of tactical warfare, they are also used to dehumanize the enemy. On the one hand, they show the power of one's own gaze; on the other, they demonstrate the complete passivity of the enemy.

With Hoskins/Shchelin (2023), the new participatory environment of war can be understood as a "crowdsourcing war". The claims, opinions, and outrage of anyone who can post, link, like or share on social media lead to the war being perceived as fragmented and thus as socially non-committal and arbitrary. Drone images in which perception, recognition and killing are superimposed in a single image space are uploaded every hour, neutralizing the individual act as an interchangeable spectacle. This is the opposite of the globalized vision of visual homogeneity that characterized the wartime era of satellite television at the end of the 20th century.

Although Telegram is not controlled by the Russian government, the presence of government or pro-government media on Telegram indicates the control of the information space by Russian propaganda. A prominent example of a channel-based Russian "war feed" is Colonel Cassad, originally a personal channel of journalist Boris Rozhin. He supports the Russian war doctrine, believes that the Cold War never ended and that the war against Ukraine is an opportunity for a 'revival' of Russia. With over 865,000 subscribers (as of July 1, 2023), Colonel Cassad has become the most important pro-Russian Telegram channel. Before the war, Rozhin was the number one communist blogger on Life Journal (colonelcassad.livejournal.com). He invented the meme "polite people" to describe the Russian units – special forces with green uniforms without insignia – that were deployed to annex Crimea.

With other Telegram channels, it is less clear who the main operator is. According to further information from Hoskins and Shchelin, one of the largest pro-Russian channels is Rybar (t.me/rybar/35537), which has 1,217,000 subscribers (as of 1 July 2023) and offers a mixture of populist infographics, tactical coordinates for possible airstrikes and criticism of the Russian Ministry of Defense. With its 'from below' perspective, it achieves a high level of credibility among the target group of the Russian armed forces.

According to the TGstat analysis system on the most popular Telegram channels in Ukraine, the Ukraine Online channel (1,350,000 subscribers as of July 1, 2023) also does not provide any open information about its operators and their interests. This channel is top-down and state-oriented, with key elements of the news feed reproducing public speeches by members of the government, bureaucratic elites, and strategically relevant information. Ukraine Online reproduces the media format of the state news channel and sees itself as a pro-government medium that primarily disseminates patriotic and pro-state content. This channel largely does not recycle authentic war-related experiences. Rather, Ukraine strategic media goal is to provide media-suitable identity scripts to build a collective memory. In comparison, the broad acceptance of Ukraine Online shows firstly that Telegram can be received as a traditionally 'serious' news channel, secondly that the agenda of cross-sector general content generates majorities, and thirdly that the war as experiential content addresses male-dominated target groups that have a lower reach in comparison.

Gamified war

In such 'male' offerings, action cams in point-of-view mode mark a new turning point in the gamification of war and violent extremism. With action-based media formats such as the livestreaming of acts of violence, selfie sticks in the trenches and action cams in field maneuvers, new practices of digital media usage in military use have emerged in the special interest channels on Telegram, which meet the demand for immediacy and realism in the online image markets.

Action cams in point-of-view (POV) mode mark a new turning point in the gamification of war and violent extremism. With action-based media formats such as live streaming of acts of violence, selfie sticks in the trenches and action cams in field maneuvers, new practices of digital media use in military applications have emerged in the special interest channels on Telegram. These practices meet the demand for immediacy and realism in the online image markets. Gamified images of war from a first-person shooter perspective are very popular on video platforms, social networks, and messaging apps. They mix images recorded on location with popular images of virtual worlds and thus blur the boundaries between reality and fictionality.

Cameras are mounted on the helmets and weapons, on armored vehicles, cannons or aircraft and show playful images of military and terrorist operations. To increase visual interest, a helmet-mounted GoPro miniature camera is used to transmit live broadcasts of war and violent extremism. Their preferred location is the scene of combat, where there is a direct encounter with the enemy/victim. For games studies scholar John Martino (2021), the images produced by both state and non-state actors (to be disseminated via digital streaming platforms and the internet) are examples of hybrid warfare.

Mobile technology and digital connectivity are essential components of the soldier's experience on the front line. Armed forces operating in the field always project themselves into a virtual script and go into battle driven by imaginary spectacles.

In the present day, the POV shot contributes significantly to the shared image repertoire and the digital visual memory of the Instagram, TikTok, Discord and

Twitch generation. The dissemination of egocentric images of the battle from the perspective of those directly involved realizes the old media dream of bringing war to life in the context of its medial observation. Moreover, such a mediatized war, live-streamed in the perspective of participation, is located at the contemporary interface of social media, providing the dialogical proximity between sender and receiver. The media dream of the immediate transmission of experience is still carried on today by the digital messengers of war. Unlike historical media systems, soldiers are willing to take more risks when they know that the currency of social media, namely 'authenticity,' demands visual credibility, or when they are aware that their images are being broadcast live on the internet. Today, soldiers are not only fighting the enemy, they are also fighting for attention on the virtual battlefield when it comes to competing for likes, comments and retweets.

The combat images on social media or messenger app platforms such as Telegram, Instagram, TikTok, Facebook and YouTube thus combine two essential elements: firstly, they ensure traditional media satisfaction through realistic depictions of war; secondly, their images create a new stage for the self-portrayal of a medially personalized war hero (equipped with selfie stick, helmet camera, direct addressing, first-person perspective) who is willing to share his personal experiences (seemingly exclusively) with his followers.

The image space is also a death zone, as it represents an instrument for recognizing the enemy and killing him with a well-aimed shot. In the center of the image space is a targeting device, a crosshair; this design offers an interface to the first-person shooter. Gamification of war and dehumanization are two sides of the same coin: the video repeats the successful killing of the enemy. Killing is portrayed as a craft that is practiced without error and without emotional involvement. The viewers in the video look through the eyes of the soldier. The camera becomes the perpetrator's companion. First the victim is observed, then the decision is made, and the viewer sees the trajectory of the bullet and the killing of the enemy. The video shows a perfect sequence of events. The person who perceives, recognizes and acts is portrayed less as an avenger and more as someone who can apply his expertise in the field.

Telegram offers an attractive interface between Agon and Ares: competition under war conditions in the presence of platform media means that military action competes for public attention. The technologies that many soldiers and civilians carry in their pockets and Telegram's information infrastructure concur in establishing a new ecology of war in which the distinction between combatants and non-combatants is blurred. When, in conjunction with clicks and likes, the relationships between friend, enemy, victim and perpetrator become excessively polarized, social media, messaging apps and online platforms contribute to the acceleration and intensification of violence.

This special issue investigates smartphone use, online media, platform politics, and the impact of the crowdsourced war. New forms of digital participation, collective witnessing and web archiving by media users and media providers are linked with new methodological and empirical challenges for source analysis of digital forensics, jurisdiction, and collective memory. The contributors analyze digital society and its relationship to war, violence, genocide, witnessing practices and cultural appropriation in a critical and reflective manner.

The issue is divided in three sub sections. The first sub section – Digital Transformations of Visual Politics and Collective Memory – comprises two articles which deals with collectively related media content.

In the article "War, Authoritarianism & Technology. The History of the Khatyn Massacre and Its Soviet and Post-Soviet Reproductions" Tatsiana Astrouskaya discusses the monument erected in commemoration of the Nazi destruction of the village and killing of the inhabitants of Khatyn in Belarus in 1943. The memory of WWII remained in the kernel of the collective remembrances in Belarus during the whole period of post-socialism, in the aftermath of the suppressed Revolution 2020, and even more so with the beginning of the Russian full-scale war in Ukraine its significance was again elevated. In this article Astrouskaya I traces how the memory of Khatyn's has been repeatedly reproduced using diverse mediums, tracing continuity and change between the Soviet and post-Soviet. She provides examples of how new digital technologies have been employed to revive and weaponize the history of WWII, and how its significantly alters the mode of the representation.

In "Izolyatsia: A Factory, But a Different One. Production of Images by the Means of Violence" Lesia Kulchynska considers how the former art center *Izolyatsia* in Donetsk has come to play a core role in the Russian war propaganda. First, the process of transformation of the art center into a prison by the Russians in 2014, where violent destructions of the art works was a core part were being filmed and spread online as propaganda. Since then, the torture rooms, like all such facilities across the Russian-occupied territories, have functioned as an extension of the Russian state-controlled media complex. Both the torture and the prisoners' confessions are being filmed and used as visual propaganda. In fact, *Izolyatsia* specialization is the production of images of the enemy, the so called "Nazis". As displayed by Kulchynska, violence, has in these cases become an instrument of image extraction, a tool for processing humans into images needed for the war machine to operate.

The second sub section – Discourse Analysis and Social Media Sensitive Subjects – comprises three articles which focuses the connection between political power and discursive order.

In "Talking about the Residents of the Territories Occupied by Russia in Official Ukrainian Discourse" Yana Lysenko explores the representation and perceptions of the residents of territories occupied and annexed by Russia since 2014 and after 2022 in the official Ukrainian discourse since the large-scale Russian invasion. It examines if and how Ukrainian official communication addresses the very sensitive subjects – the future reintegration measures concerning this population group. By analyzing statements from key formal and informal political actors within President Zelensky's government, the study employs an extended framing analysis to capture the main characteristics of the official Ukrainian discourse on the residents of occupied territories from February 2022 to December 2023. The research reveals a dominance of moral judgements and proposed military solutions over discussions on the causes of occupation or detailed plans for reintegration in the studied discourse. Findings indicate a focus on reclaiming territory, with specific attention to Crimea, and a general underrepresentation of the occupied territories' residents in the discourse. The study highlights a lack of a clear action strategy for addressing the reintegration of these residents and suggests a need for more comprehensive research on political discourse concerning the residents of the occupied territories, including broader perspectives and the reasons for their underrepresentation in official narratives.

In "Living Through Narratives. A Psycholinguistic Study of War Stories by Bohdan Lepky and Today's Ukrainians in Print and Digital Media" Serhii Zasiekin traces psycholinguistic markers of war-related trauma in the narratives of today's Ukrainians on Facebook group "Writings from the War" and of the well-known Ukrainian writer and public figure Bohdan Lepky. The aim is to identify changes as regards categories related to social relations – "We", "Social" and "Family'"– that occurred in the civilians at the end of the first year of Russia-Ukraine war and to compare these data with those found in Lepky's writings. Zasiekin display that there is greater prevalence of the categories "Affiliation" and "Achievement" in the Facebook corpus, implied that Ukrainians were praising their advances after one year of resistance and fighting against the enemy and focused psychological resilience while Lepky's stories showed none of these features implying that he could not rely on others during the tragic times of that war a century ago.

In "Strategic Deception. Russian Trolls' Attempts to Reframe the Annexation of Crimea for English-speaking Audiences on Twitter/X" Maksim Markelov analyses the temporal and contextual evolution of discursive practices of state-sponsored tweets, identified as "Russian trolls" in the context of the annexation of Crimea to Russian Federation in 2014. Markelov shed light on the discursive tactics used by "Russian trolls" leading to the full-scale attack of Russia to Ukraine in 2022. By comparing GRU (Russian military intelligence agency) actors and other users' tweets using a mix of methods he disclose two crucial under-researched aspects of the annexation: the annexation as a pivotal moment that reshaped Russia's social media landscape and Russia's turbulent societal and social media environment after the annexation, which created fertile ground for language change.

The third and final sub section – Digital Warfare. Strategies and Tactics – comprises two articles which considers actor-centered practices that make war perceived as both a power-political and an empowering and subversive intervention.

In "A Spinozist Approach to Political Agency Online" Max Kramer discusses the war-like conditions online and the meaning of the strategies and tactics used by various actors. He argues, with far-right populists in India as case, how strategies and tactics online can be better understood from a Spinozist perspective. In this contribution Kramer addresses various studies that differentiate between tactics and strategies to comment on how agency on social networking sites (SNS) is exercised and conditioned arguing that a theorization of the political is necessary to locate the adequate place for concepts of strategies and tactics. I argue that in many explorations of the concepts of "strategies" and "tactics" there are underlying assumptions about what kind of space the political is that participate in a martial impasse regarding SNS. Furthermore, this impasse is itself driven by the antagonistic qualities of SNS and the lack of institutional regulation. I argue that Baruch de Spinoza's political ethics of *ratio* and his relational ontology provide ways out of this impasse.

In "When the war goes viral. Ukrainian and Russian War Memes" Elena Korowin analyses memes which became one of the most powerful weapons in 2022 in Ukraine's information war against Russia, both produced in Ukraine and within Russia. As Korowin displays through a iconographic review of a number of different visual and textual themes a characteristic of these memes are their self-empowerment through humor and easy to get appeal. Korowin argues that such political memes have surged in the recent past and that they serve the same function as posters, flyers and signs held up at demonstrations, only their reach and distribution can be much more effective as they are distributed via the Internet.

References

Ford, Matthew, and Andrew Hoskins. *Radical war: Data, attention and control in the twenty-first century.* Oxford University Press, 2022.

Gunning, Tom. "The cinema of attraction [s]: Early film, its spectator and the avant-garde." In: *Theater and Film: A Comparative Anthology* 39 (1986).

Hoskins, Andrew, and Pavel Shchelin, "The war feed: Digital war in plain sight." In: *American behavioral scientist* 67.3 (2023): 449-463.

Martino, John. *Drumbeat: New media and the radicalization and militarization of young people.* Routledge, 2021.

Merrin, William. *Digital war: A critical introduction.* Routledge, 2018.

Sontag, Susan. *Regarding the pain of others.* Macmillan, 2004.

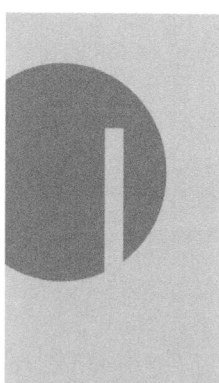

**Digital Transformations
of Visual Politics
and Collective Memory**

War, Authoritarianism & Technology
The History of the Khatyn Massacre and Its Soviet and Post-Soviet Reproductions

Tatsiana Astrouskaya

Abstract

On March 22, 1943, the village of Khatyn was burnt down together with its 149 inhabitants by the Nazis and their collaborators. In 1969, a magnificent monument was erected on the site of the extermination, commemorating the suffering and resilience of Khatyn inhabitants quite in the spirit of Brezhnev's memory policies of the time. The monument attracted millions of local and foreign tourists, including dozens of high-level officials, while its images were reproduced in countless copies in the Soviet Union and abroad.

The village's history and its various reproductions rendered Belarus' particularity, laying the foundation of its post-war identity, Soviet and post-Soviet alike. Not less significant was the "export" value of Khatyn, its place as a major tourist attraction – the painful experience of war and its overcoming were melted together with the success of the Soviet modernization, and, ultimately, ensured the country's (even if limited) political autonomy. The memory of WWII remained in the kernel of the collective remembrances in Belarus during the whole period of post-socialism, in the aftermath of the suppressed Revolution 2020, and even more so with the beginning of the Russian full-scale war in Ukraine its significance was again elevated.

Aliaksandr Lukashenka's authoritarian regime announced the massive reconstruction of the Khatyn memory site, adjoined by an ambitious media campaign. In search for legitimacy, the regime embarks on the symbolic meaning of Khatyn, attempting to claim its geopolitical significance in the region, justify its animosity to Ukraine, and patch internal raptures in the Belarusian society. In this paper, I trace how the memory of Khatyn's has been repeatedly reproduced using diverse mediums, tracing continuity and change between the Soviet and post-Soviet. I am interested in how new digital technologies have been employed to revive and weaponize the history of WWII, and whether the former significantly alters the mode of the representation of the latter.

Keywords

Authoritarianism; post-war identity; Soviet era; post-socialism; media politics

Introduction: War as a Perpetual Performance

On March 22 1943 the combatants of the SS Commandos 118 and Dirlewanger murdered 149 inhabitants of a small Belarusian village Khatyn'(Хатынь). Some were shot, but most were hunted down, driven together in a barn, and burned alive. It is believed that this punitive action was a response to a partisan combat in which several German soldiers and officers were killed (Rudling 2012: 38). Almost the whole population of Khatyn perished in this killing, only three children and one adult man survived. Located some 60 kilometers North of the Belarusian capital of Minsk, the Belarusian counterpart of Oradour-sur-Glane was never rebuilt. It disappeared from political maps to reappear, decades later, in the mental maps of WWII East European memory entanglements.

Being unspeakably terrifying this massacre was only one episode of the hundreds and thousands of similar tragedies during WWII and the subsequent wars and could have sunk into oblivion, as many others of this kind.[1] And yet, the small village of Khatyn turned out to be at the epicenter of an extraordinary fiddling with historical narratives that continues up to the present day albeit with new digital means. The story of Khatyn village and its inhabitants formed the foundation of the Belarusian Soviet and post-Soviet memory politics, ultimately, under new digital conditions, as I will demonstrate in this article, substituting for the latter.

Rapid digitalization significantly altered our relationship with the past (Fickers 2022: 50), and authoritarian regimes increasingly use this change to their advantage (Schlumberger et al. 2023). Also, the ongoing war in Eastern Europe and utilization of digital technologies in this war (Ford&Hoskins 2022) has intensified the attempts to revoke history for the sake of current political purposes that the populist authoritarian regimes of Aliaksandar Lukashenka and Vladimir Putin have practiced for decades (Rudling 2008, Goujon 2010, Marples 2014, Klymenko 2016, Edele 2017, Kurilla 2020, Weiss-Wendt 2020). Similarly, if entertaining its subjects has been inherent in all political regimes, with the advent of technology, such as television (Bourdieu 1998: 21) and social media the place of the visual

---

1 In Belarus alone, the number of villages burned with their inhabitants is close to seven hundred. Altogether some ten thousand Belarusian villages perished in a fire during WWII. Also, in these cases human casualties were high. The State Prosecutor's Office in Belarus now continues the investigation, bringing in new cases and claiming that these numbers were even higher. The results of this investigation, however, cannot be independently verified at the moment. Even though the Nazi war crimes in Belarus indeed require additional investigation and reflection, the corruption of the Judiciary System in Belarus and the continuous weaponization of history by the regime of Lukashenka gives many reasons to doubt the objectivity of the investigation.

(Makarychev 2021, c.f. Kazharski 2023) and the role of entertainment in information production and dissemination rises exponentially. Post-Soviet authoritarian regimes utilize these instruments in an increasingly effective way (c.f. Soldatov and Borogan 2015), even if remaining intellectually rooted in the conservative ideology of the Brezhnev era, the Cold War East-West confrontation, and Soviet memory politics. In this vein, the memory of war has been revoked not only to mobilize but also to amuse (c.f. Gabowitsch 2020). This blending of technological innovation and anachronistic thinking, of horror and pleasures of war representations will be of the interests of this article.

The idea that the Soviet ideology continues to provide sustenance for Putin's and Lukashenkas's authoritarian regimes has been already commonplace, yet what are the exact mechanisms and instruments of such constant reproduction of and the reference to the Soviet past? How do they correlate with the initial event, and in which relationship Soviet and post-soviet interpretations of this event find themselves? Digitalization intensified the discussions about how past and present, truth and lies, copy and original (of meaning, image, object, event) interflow and mix so that it becomes difficult not to say impossible to differentiate one from the other. "Digital time is an indefinite present […]" – commented historian Marcello Ravveduto (Ravveduto, 2022:135). A digital copy is an infinite original one may add.

Digitalization enabled dictatorships with the instruments that automatized the radical detachment from truth Hannah Arendt warned us about (Arendt 1971). Similarly, it paradoxically aggravated their symbolic dependence on what has been represented as original and trustworthy. Technologically advancing in deceiving, faking, forging, and falsifying, the autocratic regimes continue to fiercely claim the authenticity of their (imagined) foundation. Wars, past and present, being in many respects, an ultimate human experience, offer a fertile soil for such claims.

In this article, I demonstrate, how this happens, tackling one particular war event – the brutal extermination of Khatyn village and the history of its continuous reproductions by Soviet and post-Soviet regimes in Belarus that (with brief interruptions) had lasted already for more than half of the century. I look into the continuity and change of this performance from a longue durée perspective and through the prism of contemporary technical development.

I am asking, by which means and mediums – visual and textual – (digital) dictatorships aim to achieve an effect of a continuous, never-ending presence of the past that interferes with and shapes the present, calling for retribution and revenge in the present wars. And finally, how digital reproducing and performing interfere with the both initial event and its Soviet reproduction that has been referred to, treated, and celebrated as "original".

Khatyn: A Short History of a Longstanding Brand

Some twenty years after the end of WWII, the Khatyn massacre started to transform from one of the most documented war crimes in the Nazi-occupied territories of Eastern Europe to, ultimately, a well-recognizable brand, that Soviet and post-Soviet authorities successfully promoted both inside and outside of the country.

What were the reasons for such metamorphosis? The rise of attention to the history of Khatyn resulted from the shift in Soviet memory politics with the new Tsk KPSS First Secretary Leonid Brezhnev's 1964 elevation to power. Singling out WWII (or, how it was and continues to be called, the Great Patriotic War, 1941-1945) as a foremost event of Soviet history, the new Party leadership launched the beginning of the colossal commemoration campaign. This included pompous celebrations of Victory Day – on the 9th of May, and the erection of grand war monuments across the Soviet Union (c.f. Oushakine 2013: 289-290, Gabowitsch 2020).

In Soviet Belarus, Petar Masheraŭ a former partisan, and a decorated war hero became the First Secretary and de facto the head of the republic in 1965. Masheraŭ's vision of Belarus hit worst of all the Soviet republics by the war, was to a significant extent shaped by the ultimate experience of this catastrophe. "For Soviet people" – proclaimed Masheraŭ opening the all-Union writers' assembly (1975) in Minsk – "the fiery mileposts of war never belonged and will never belong to history".[2]

The heavy war devastations made way for the astonishingly prompt and effective post-war restoration, generously supported by the all-Union budget. Thus, post-war Soviet Belarus became one of the most successful showcases of socialist modernization, and Masheraŭ was among its main architects. Blending the Soviet success and the tragic war experience, he could reinforce the Soviet Belarusian identity, as it happened with the iconic brand of Belarus as the "heroic partisan republic" (Rudling 2008, Lewis 2019, Marples 2014, Chernyshova 2023). The story and memory of Khatyn became another powerful element of the post-war Belarusian identity.

---

2   The assembly was devoted to the representation of the Soviet people's heroism in literary works. Vystuplenie 1-go sekretaria TsK KPB P.M. Masherova na Vsesoiuznom soveshchanii pisatelei i kritikov v g. Minske, posviashchennom otrazheniiu v sovetskoi literature podviga naroda v Velikoi otechestvennoi voine. 1975, https://archives.gov.by/home/tematicheskie-razrabotki-arhivnyh-dokumentov-i-bazy-dannyh/vydayushhiesya-lichnosti/arhivnye-dokumenty-i-materialy-3/uchastniki-partizanskogo-dvizheniya-v-belarusi-rukovoditeli-bssr-i-sssr-v-50-70-e-gg-hh-v-2/masherov-pyotr-mironovich/vystuplenie-1-go-sekretarya-czk-kpb-p-m-masherova-na-vsesoyuznom-soveshhanii-pisatelej-i-kritikov-v-g-minske.

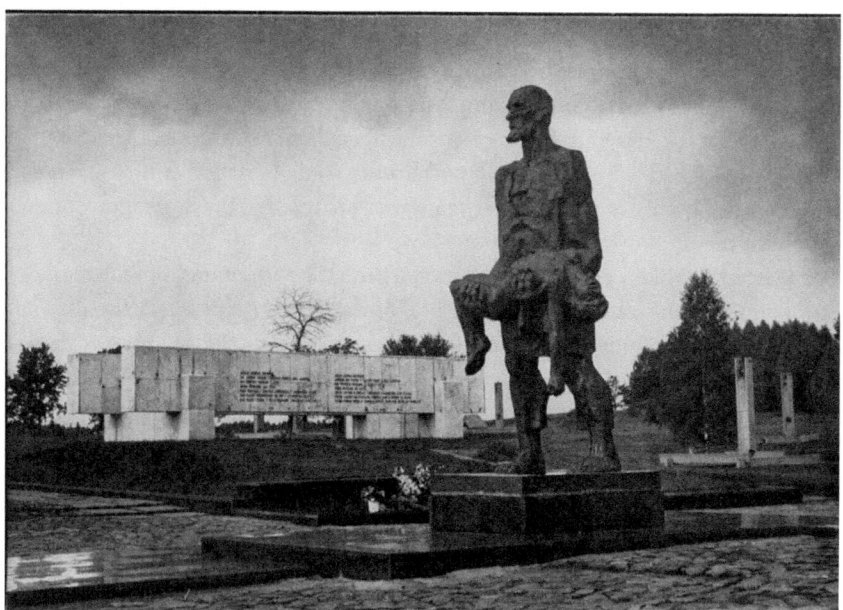

Figure 1: The Memorial of Khatyn, 1968-1969 (architects Yu. Hradaŭ, V. Zankovich, L. Levin, sculptor S. Selikhanaŭ) and its central element – the sculpture of "An Unbowed Man". Source: Khatyn. The Memorial Complex. Aurora Publishers, Leningrad. Private archive of the author.

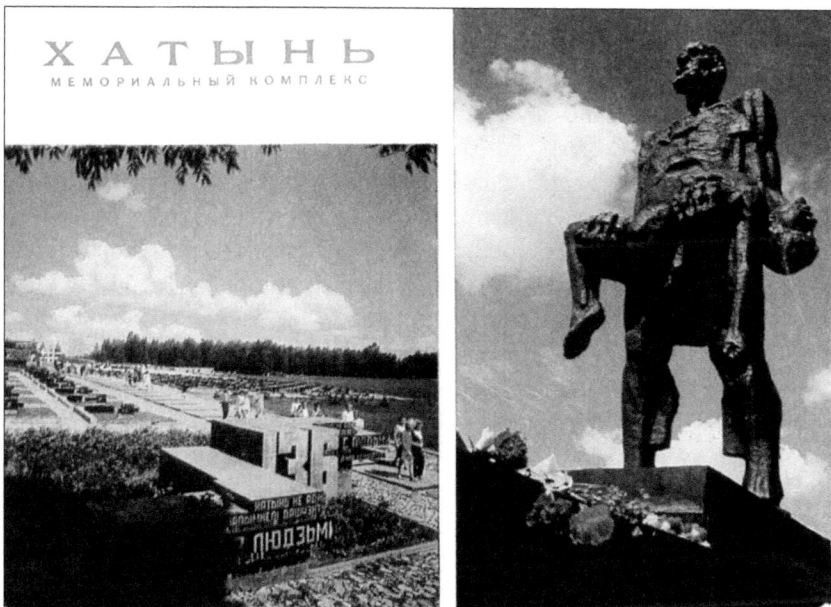

Figure 2: The Memorial of Khatyn. Fragments. A postcard issued by the USSR Ministry of Communications (1975). Photo by B. Elina. Private archive of the author.

The eponymous monument of Khatyn (Fig. 1, 2) was erected on the site of the executed village in 1968-1969 (architects Yu. Hradaŭ, V. Zankovich, L. Levin, sculptor S. Selikhanaŭ) and became one of the most recognizable symbols of Belarus and a popular tourist attraction and since then. As anthropologist Serguei Oushakine argued "The Khatyn Memorial was an emblematic example of a quick and remarkably successful memorialization campaign that the Soviet government started in the mid-1960s." (Oushakine 2013: 289).

Khatyn was likely chosen and promoted to distract attention from the memory of *Katyn* (Катынь), one of the most brutal Soviet crime scenes near the Russian town of Smolensk, where thousands of Polish citizens, including many officers, were executed in 1939-1940. The similarity of names should have confused the international audience concerning who was the perpetrator in Katyn (Lüer 2002).[3]

Similarly, the brand of Khatyn offered an opportune counter-narrative to the memory of the Holocaust. According to different estimations 2.1 to 2.5 million people including 800,000 local and European Jews were murdered in the territory of Belarus during WWII. Soviet authorities disguised the Nazi's genocide of particular ethnic and social groups, including Jews, Cinti, and Roma, insisting on the homogeneity of the Soviet experience of suffering and resistance. A Belarusian village was a suited place to showcase the victimhood and heroism of the *Soviet* people (Lewis 2015: 371). Jews mostly inhabiting cities and small towns (Stetlekh) by the time of the Khatyn Massacre (March 1943) perished in the Holocaust. It is noteworthy that the Jewish movement for emigration from the USSR gained momentum at the very time when the commemoration of Khatyn launched. Among other questions, the emigration triggered the discussion of the Holocaust and its disavowal in the Soviet Union with the new force – both within the country and internationally.

Simultaneously, Khatyn offered more than a Soviet counter-narrative to the inconvenient past. The two identities of Soviet Belarus as the pioneer of successful socialist modernization and urbanization and the utmost victim of WWII paradoxically amalgamated in how this story was represented and disseminated. One of the most telling visualizations of this amalgamation is the page of the "Travel to the USSR" magazine (a Soviet outlet distributed in the West to propagate the Soviet way of life and motivate Western tourism to the USSR) (Fig. 3). The page entitled "I saw Byelorussia"[4] depicts the "Bells of Khatyn", a part of the monument on the top of the page, and happy, just married couples strolling across the promenade of the Svislach river in the center of Minsk, laughing and chatting livingly.

---

3   The Putin regime today attempts to reverse the story once again, shifting the blame at Germans (Kragh 2024).

4   The name the republic that was often used in the contemporary English-language publications derived from the Russian version of the name.

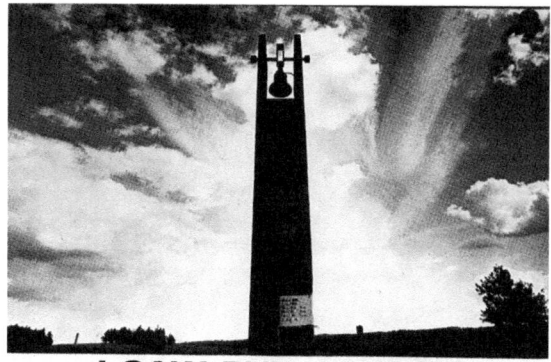
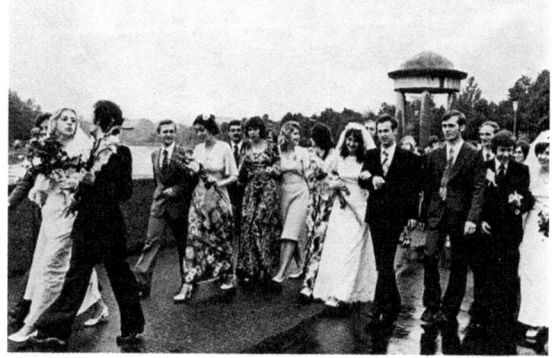

Figure 3: A back page of the "Travel to the USSR" magazine inviting Western tourists to visit Soviet Belarus. The page features the Khatyn Monument at the top of the page and a happy life in contemporary Soviet Belarus at the bottom as if linking past and future together. Source: Travel to the USSR 6 (1978), Archive of the Francis Skaryna Belarusian Library and Museum, London, UK.

The attention to Khatyn was constant during the Soviet time. In the 1990s it dropped massively, and then, in the early 2000s raised again. This was the moment, when Aliaksandr Lukashenka, the winner of the first and only democratic presental elections in Belarus in 1994 strengthened his position in power and established an authoritarian rule. Following his (as it turned out) temporary divorce from Russia and Vladimir Putin, Lukashenka started to invest much effort in memory politics. Within the next decades, the Lukashenka authoritarian regime would make the revisions of historical narrative to one of its foundation stones.

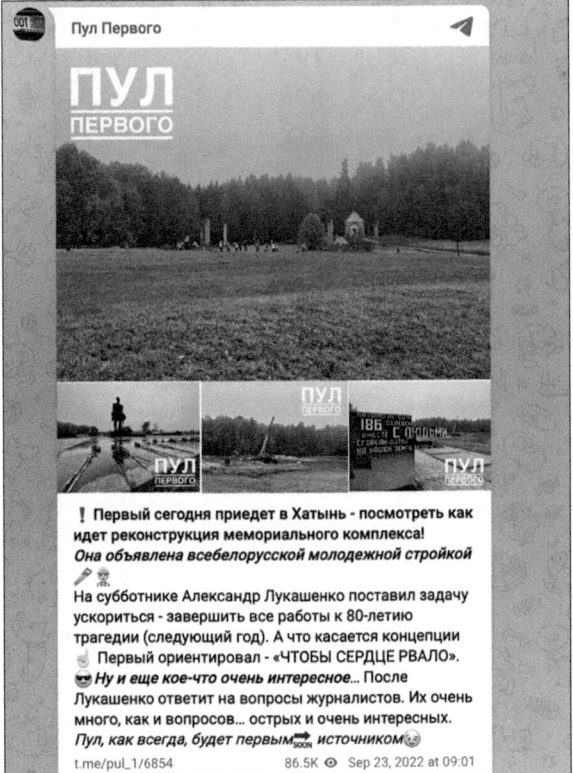

Figure 4: Reconstruction works in Khatyn as reported about by the close to Lukashenka Telegram channel "Pul Pervogo". It is reported that Lukashenka himself closely supervised the reconstruction works. Source: @pul_1 (September 23, 2022). Reproduced in accordance with Belarusian copyright provisions.

The story and monument of Khatyn regained their symbolic value from the very beginning (Rudling 2012: 45-46), yet their meaning for the regime grew drastically in the aftermath of the Belarusian Revolution of 2020 and the full-scale Russia's war in Ukraine in 2022 (Ackerman 2021, Astrouskaya 2022: 25-26). The Memorial was massively (and urgently) reconstructed in 2023 on the occasion of the 80th anniversary of the massacre, and a new museum and an Orthodox Church were built on the site in a record-breaking time (Muzei, 2023). Ultimately, Khatyn became *the* place of memory around which the narration of Belarusian history has been now entwined. Simultaneously, from 2019 the messenger Telegram has increasingly become *the* medium, through which the quickly revised narratives of Khatyn have been produced and disseminated (Asrouskaya 2022).[5]

The reconstruction works, regularly spring cleans and Soviet-style subbotniks, are necessarily lived-streaming and accompanied by detailed photo- and video reports in media, particularly on messenger Telegram (Fig. 4)[6].

A War Crime, a Monument, an Image, an Icon

As early as 1942 Soviet authorities started to document and investigate the Nazi crimes in the occupied territories, and these efforts continued well into the 1980s (Exeler 2016, 824). Their special attention attracted also local (Belarusian, Ukrainian, Lithuanian, Latvian) collaborationist units. The Soviet investigators of the Khatyn Massacre documented it as early as 1944 (Rudling 12, 29). Nevertheless, in the official memory and the public discourse, the story of Khatyn did not exist until the late 1960s.

The electronic catalogue of the National of Belarus, possessing the largest and the most comprehensive library collection in Belarus, allows tracking of the quantitative and qualitative change of publications on Khatyn over the whole post-war period. The numbers were manually extracted from the catalogue using various search filters and then processed with Excel PivotTables. The following graph (Fig. 5) displays the change in publication numbers perhaps in a most salient way. The number of printed books, and audio and visual materials on Khatyn, was equal to zero between 1943, the year of the massacre, and 1968. After 1969 the number of publications rocketed, which supports the presumption about the centrality of the Khatyn story. The substantial growth of publications from 2002 to 2022 justifies the assumption about the attempts to instrumentalize the image of Khatyn by Lukashenko's regime. For the most part, these materials were intended for a broader public, but there were scientific studies and document collections.

---

[5] The role of the Telegram messenger in consolidation of the Belarusian protests 2020 has been already discussed in detail, yet, also the Belarusian regime takes advantage of the liberal anti-censorship politics of this social media and increasingly uses Telegram to spread its influence (c.f. Wijermars&Lokot 2022: 126, Astrouskaya 2022: 24-26, Koran 2023: 72). See also a more detailed analysis on the role of Telegram below.

[6] Here and passim. Working on this article I have created a collection of Telegram entries devoted to Khatyn between 2019 and 2023, they can be consulted via the public Conifer collection "PerformingKhatyn" via https://conifer.rhizome.org/astrouskaya/performingkhatyn.

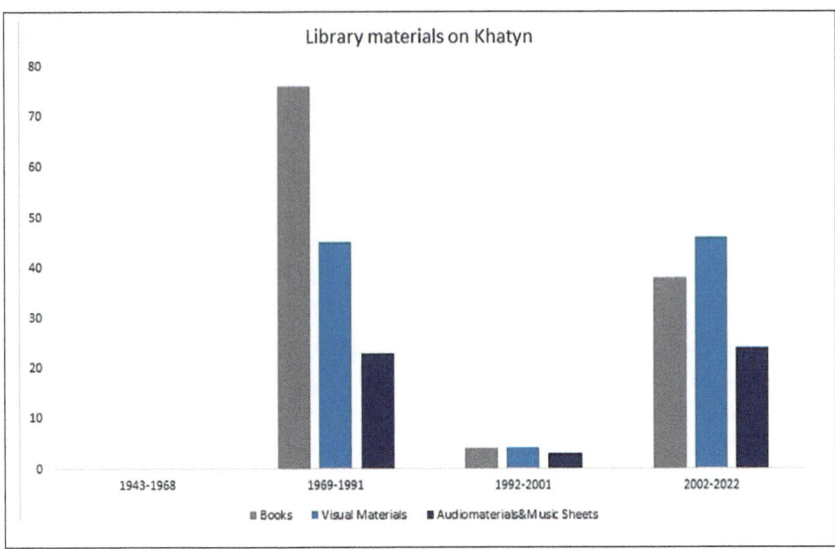

Figure 5: The number of publications on the Khatyn was issued between 1943 and 2022. Source for Data: Electronic Catalogue of the National of Belarus, https://e-catalog.nlb.by. Data collection&analysis: Tatsiana Astrouskaya, data processing: Oleg Sidarenka.

Peculiarly, the new memory politics in the Soviet Union and the boom of technology were tightly bound together. Already Friedrich and Brezinski argued that the new means of mass communication have significantly reinforced the ways authoritarian regimes spread their ideology and influence (1965: 130). These new technical means were intensively engaged also in the Khatyn commemoration campaign. Booklets, tourist maps, postcard collections, and pins were manufactured, printed, and circulated *en masse* from 1969 on. The massive inflow of visual material had contributed to the recognizability of Khatyn. Booklets and journals aimed at Western audiences featured Khatyn along with a successful modernization story as *the* symbol of Soviet Belarus. The "partisan" brand was created in the first place for internal, Soviet use, in order to bargain a certain autonomy for the republic within the hierarchy of the Soviet "family of nations." With the brand of Khatyn, Soviet image-makers clearly aimed to reach an audience outside of the Soviet Union (Fig. 6).

The following graph (Fig. 7) visualizes languages of published material and their share in the overall publications list on Khatyn since 1969, according to the electronic catalogue of the National Library of Belarus, demonstrating the intent at international outreach. The Story of Khatyn was told and disseminated in various languages. While the Russian and Belarusian languages dominate, English, French, and Spanish also rank proportionally high making their notable share in the overall number of publications. While seen in the correlation with previous results, the language of the publication graph also displays the fluctuations of the Soviet and Post-Soviet state interest in Khatyn, with the significant growth during the period of late socialism and the returning interest during Lukashenka's embarkment on the history politics since 2001 and the history of WWII in particular.

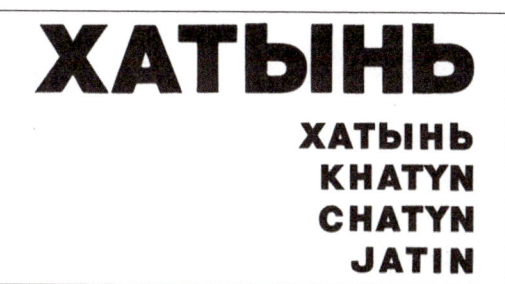

Figure 6: A flyleaf of the multi-language album devoted to the story of Khatyn and the Khatyn Memorial in Russian (Belarusian), English, German, and French languages. Source: Kaziulia Iauhen, Vetsik, Mikhail, Traianouski Ales. Khatyn. Minsk: Belarus, 1977.

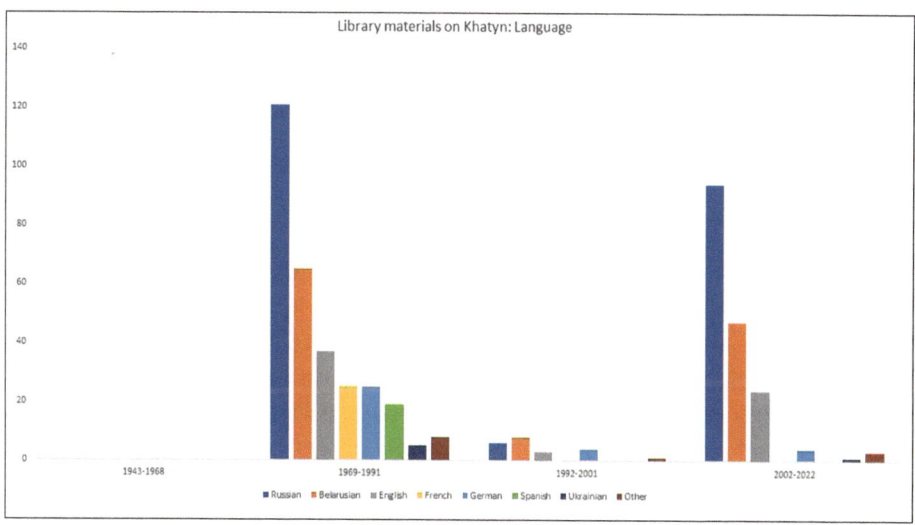

Figure 7: The language of publications on Khatyn. Source for Data: Electronic Catalogue of the National of Belarus, https://e-catalog.nlb.by. Data collection&analysis: Tatsiana Astrouskaya, data processing: Oleg Sidarenka.

It is worth noting though, that materials were for the most part printed and published in Minsk (Fig. 8). And even if Moscow was indicated as a place of publication, most of the preparation work had been done on the ground. This allows assuming that even if supported in Moscow impulse did come also from Minsk and the authorities of Soviet Belarus.

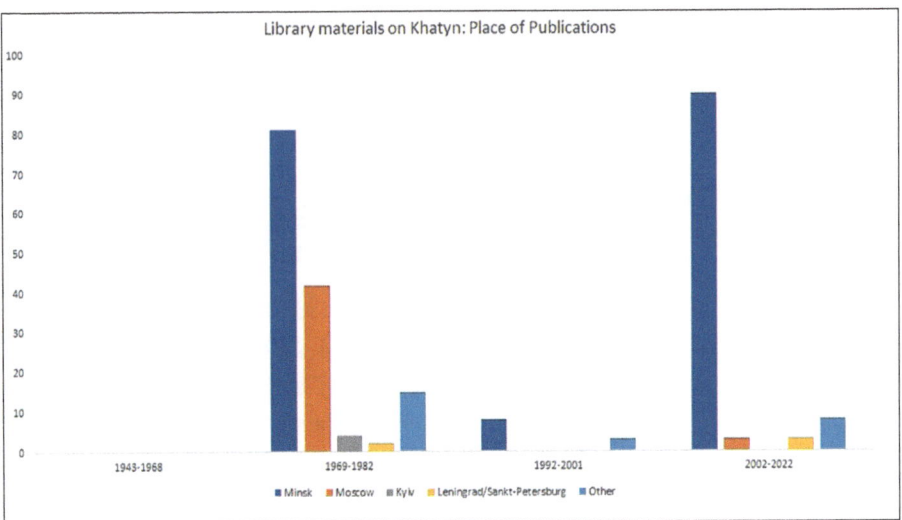

Figure 8: The place of issuing, the publications on Khatyn. Source for Data: Electronic Catalogue of the National of Belarus, https://e-catalog.nlb.by. Data collection&analysis: Tatsiana Astrouskaya, data processing: Oleg Sidarenka.

The relative rise of welfare in the Soviet Union enabled the private ownership of audio and video recording devices. This in turn supported the commemoration efforts of the state from below. Motivated and organized top-down by local trade unions or Komsomol organizations visiting tours to Khatyn (see more below), engaged Soviet citizens into a closer interaction with the memory site, often mediated by photo or (less often) video camera. However, the results sometimes were quite divergent from what the Communist Party Propaganda and Ideology Departments envisioned. The most prominent example is the work of three Belarusian intellectuals – Ales Adamovich, Ianka Bryl, and Uladzimir Kalesnik who traveled across Belarus with a microphone, recording apparat and a camera, taping the voiced and capturing the faces of the "fiery villages" survivors. This how they called the villages that suffered a fate similar to that of Khatyn, who were burned to the ground with their inhabitants (altogether 627 to their knowledge)[7]. The result of the expedition was a documentary "The Fifth Killometr", the book "I am From a Fiery Village" (Adamovich, Bryl and Kalesnik 1975) (Fig. 9), and the vinyl plates featuring the voices of survivals. Simultaneously, the collected reminiscences did not let Adamovich go for many years. He also attempted to rework them literary in his internationally acclaimed "Khatyn" novel. Some years later he also wrote a scenario for Elem Klimov's "Idi and Smotri" (Go and See, 1985), one

---

7   Belarusian authorities claim that they recently discovered more of such cases, but these claims cannot be independently proven at the present moment.

of the most award-winning films in Soviet history that nevertheless was for a great while censored by Soviet authorities before release.

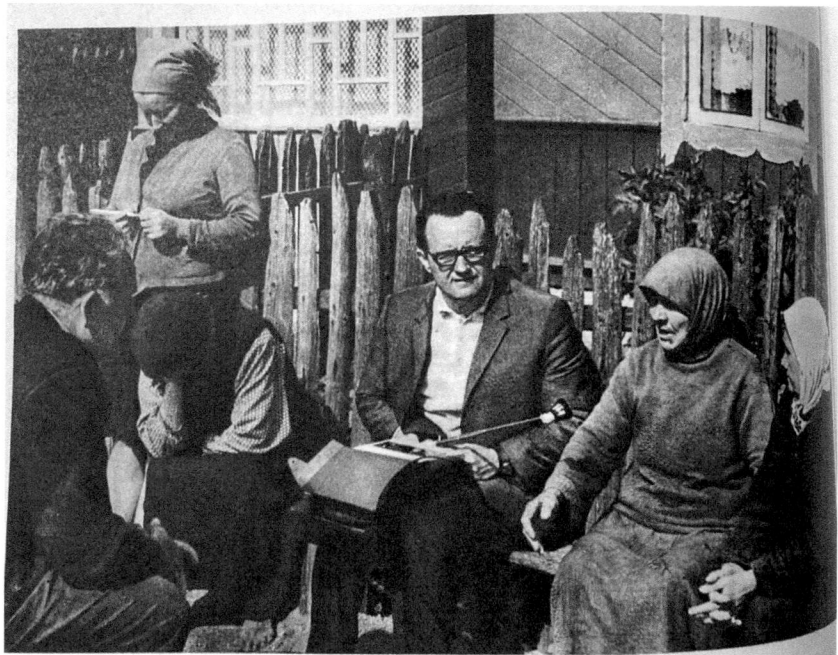

Figure 9: A Photo of Ales Adamovich taping the eyewitnesses' reports with the help of a hand tape recorder. Source: "I am from a Fiery Village" by Adamovich, Bryl, and Kalesnik.

The use of technical means in a literary and cultural rethinking of the war history was innovative; it allowed to connect participants and observers, introducing the latter to the direct experiences of the war. At the moment when the power of communicative memory slowly faded away the technological advancement allowed both spreading propaganda more effectively and diversifying the centralized interoperations offered by the state. As in the case of other ambivalent roots of everyday behaviour described by Michel de Certau (1984: xxiv), the story of Khatyn, imposed from above, yet often eagerly internalized from below took a firm root in popular culture (c.f. Gabowitsch 2020). The Lukashenka regime used this repute of Khatyn to its own advantage in its struggle for influence in the Belarusian society, and, especially, during the past years.

Significantly, it is the rise of social media especially of the messenger Telegram and their increasingly intensive use by the Lukashenka regime that got the Khatyn story a new lease on life. Lukashenka recognized the crucial role of the internet already in 2010, when the role of digital technologies in the consolidation of protest became apparent after the third presidential election in 2010, subsequent protests, and their violent suppression. This coincided with the Arab Spring and the role social media played in the mass revolutionary movement across world (c.f. Wolfsfeld, Segev, & Sheafer 2013; Markham 2014). Since then, he attempted to meander between potentially high gains of the technological innovation for the economy of Belarus and the similarly high risks of society's democratization that the development and enhancement of internet infrastructure entailed. Since 2010 and especially in 2019, the regime increasingly used digital tools for control and surveillance, and it started to act offensively using new media technologies, including social media and the Telegram messenger, so we have every reason to think of the Belarusian regime as a "digital" or at least "digitalizing" dictatorship. While traditional media usually lacking resources and having only limited coverage traditionally lag behind the Russian media sphere, the messenger Telegram turned out to offer convenient easily accessible tools of media presence not only for grassroots activists but also for "small" digital dictatorships such as that of Lukashenka's.

Since 2020, there has been a massive media campaign on Khatyn taking place in Belarus, the main instrument of which is the social messenger Telegram. While the commemoration events do take place on the spot, it is their representation on the Internet that seems to matter. The most magnificent celebrations were organized in 2021, as a reaction of the regime to the threats posed by the massive post-electoral protest of 2020, when protesters were (and continue to be) compared to the Nazi perpetrators and the rhetoric of "genocide" committed against the Belarusian people (c.f. Ackermann 2021: 2-3) was running high on the state-sponsored media and social media channels. The main pro-regime channel @pul_pervogo advertised "All-Belarusian Prayer" as a central part of the official commemoration campaign (Fig. 10, top left). Although being organized on-site, the prayer was broadcast widely. Not specifically referring to any of the major confessions in Belarus (Orthodoxy, Catholicism, Judaism, Islam)[8] and paradoxically filled with an utterly profane meaning, the event and its representations on media intensively used religious connotations to accumulate their resources. Already at this point, Khatyn had been represented as the major event in Belarusian history that embodied "our duty, our memory, [and] our history." (Fig. 10, bottom left). Another colossal commemoration campaign formally dedicated to the eightieth anniversary of the Khatyn massacre was staged in 2023, when the war in Ukraine was already in full swing and the Lukashenka regime had dragged Belarus into this war (Fig. 10, right).

---

8   The attempts of regime to Khatyn using the symbolic value of religion can be tracked to 2018 and the 75th anniversary of the massacre, when the representatives of different confessions were brought together to the memory site for the first time.

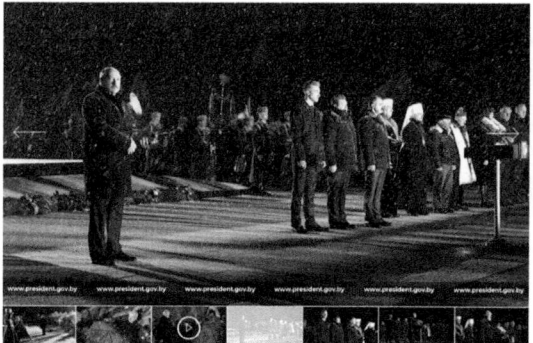
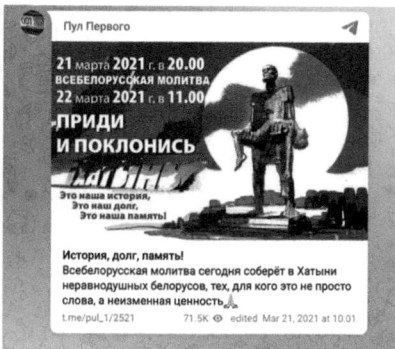
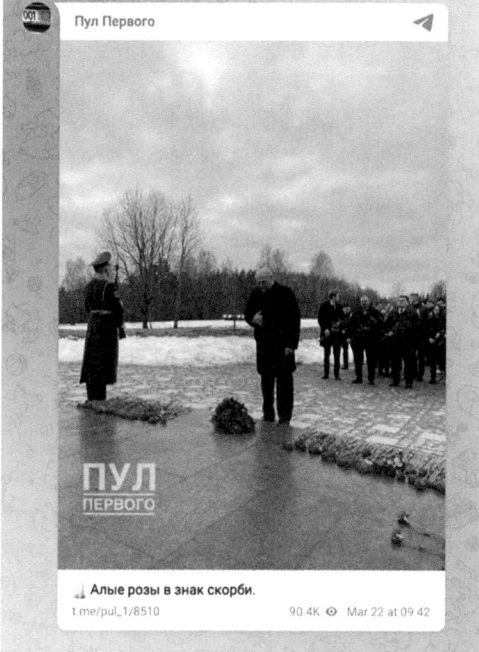

Figure 10: Commemoration events in Khatyn in 2021 (top) and 2023 (bottom). Source: https://president.gov.by/ru/events/uchastie-v-respublikanskom-mitinge-rekvieme-k-godovshchine-hatynskoy-tragedii (March 21, 2021) and @pul_1 (March 21, 2021 and March 22, 2023). Reproduced in accordance with Belarusian copyright provisions.

Major Tourist Attraction

The Soviet memorialization of Khatyn coincided with the decision of the Soviet state to prompt local and international tourism across the Soviet Union, Belarus was no exception. Khatyn, in turn, became one of the key tourist destinations in Soviet Belarus – two million people a year visited the Memorial in 1988, and the total number of visitors between 1969 and 1993 amounted to 33 million (Oushakine 2013: 293) (Fig. 11).

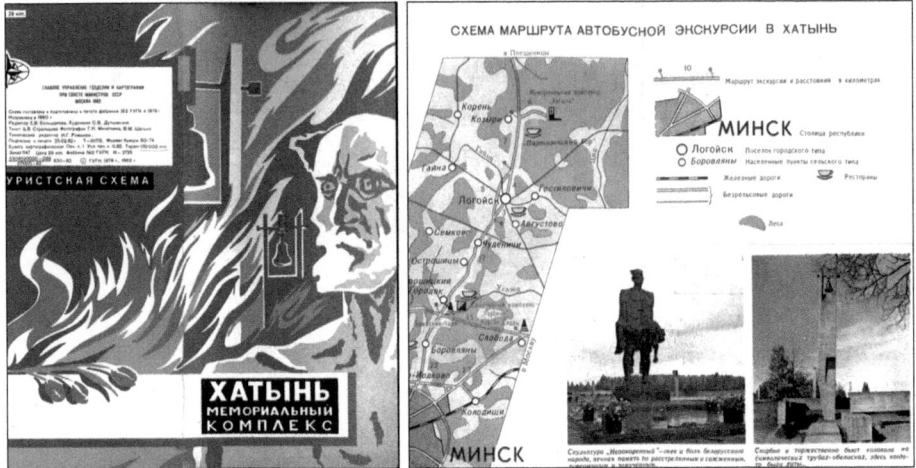

Figure 11: A tourist map. Khatyn. Memorial complex. Featuring a memorial and the route from the capital city of Minsk to Khatyn. Produced in 1982 (first edition in 1979) the Factory N2 of the Head office of the geodesy and cartography of the SSR Ministry of Geology subordinated directly to the Council of Ministers of the USSR. Personal Archive of the author.

Years after years, visitors from all over the Eastern Block, foreign government delegations, touring artists ensembles, and local tourist groups felt obliged to stop by the Khatyn memorial. One of the early visitors was that of US President Richard Nixon, during his USSR trip in the Summer of 1974. An iconic photo of Nixon (Fig. 12) sitting all alone at the varnished wood table strikingly contrasts the contemporary images of commemoration events in Khatyn, circulated in great numbers by the pro-governmental Belarusian media. The emptiness, the silence, and the loneliness of the place at appears in Nixon's photo made way for flashy colors of garish delegations of governmental officials, bank clerks, and border patrol units (Fig. 13) brought to Khatyn on each anniversary of the tragedy.

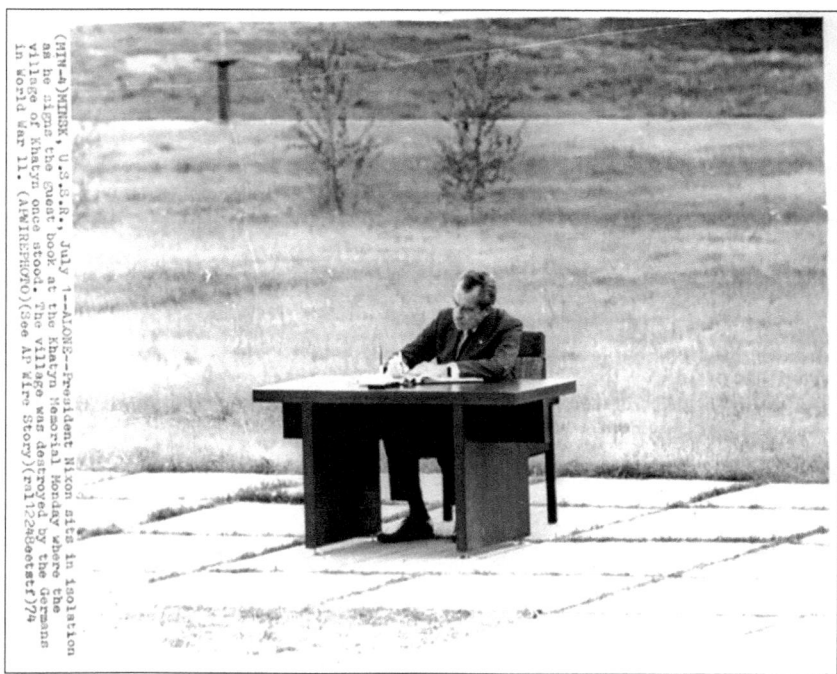

Figure 12: The US President Richard Nixon signed the guest book during his visit to the Khatyn Memorial on July 1, 1974. Official press photo. Archive of the author.

Showcasing visitor numbers seems to be similarly important for present Belarusian authorities. In 2023, the pro-governmental Telegram channel @sbbytoday regularly informed its readers about the growing number of visitors. In October 2023 the overall number of visitors reported for the current year was 450 thousand (including 115 thousand visiting the newly erected museum)[9]. The guests arrived from more than 30 foreign countries, including "Nicaragua, Israel, Myanmar, Italy, India, Egypt, Iran, Kenia, China, Cuba and others."[10] In February 2024 it was first reported about more than 520 thousand (including 11 thousand 'foreign tourists')[11] and then more than 500 thousand visitors for the previous.[12] When compared to the Soviet statistics, these numbers raise questions about their reliability. As a closed country, today's Belarus is hardly in a position to compete with organized Soviet mass tourism.

9   "V 2023 godu kompleks 'Khatyn' posetili pochti 450 tys. chelovek," October 20, 2023 (https://t.me/sbbytoday/99318).
10 Ibid.
11 "Boleee 11 tysiach inostrannykh turistove posetili Khatyn'v 2023 godu," February 7, 2024 (https://t.me/sbbytoday/108993)
12 "Bolee 500 tysiach chelovek posetili memorial'nyi kompleks 'Khatyn'' v 2023," February 29, 2024 (https://t.me/sbbytoday/110979)

Who visited Khatyn was also important. In the Soviet Union – foreign delegations both from the friendly socialists and Western countries traveling to Moscow and entering the Soviet Union through its Western border, the Belarusian city of Brest[13] was necessarily brought to Khatyn, initially having no idea about its history (Fig. 14).

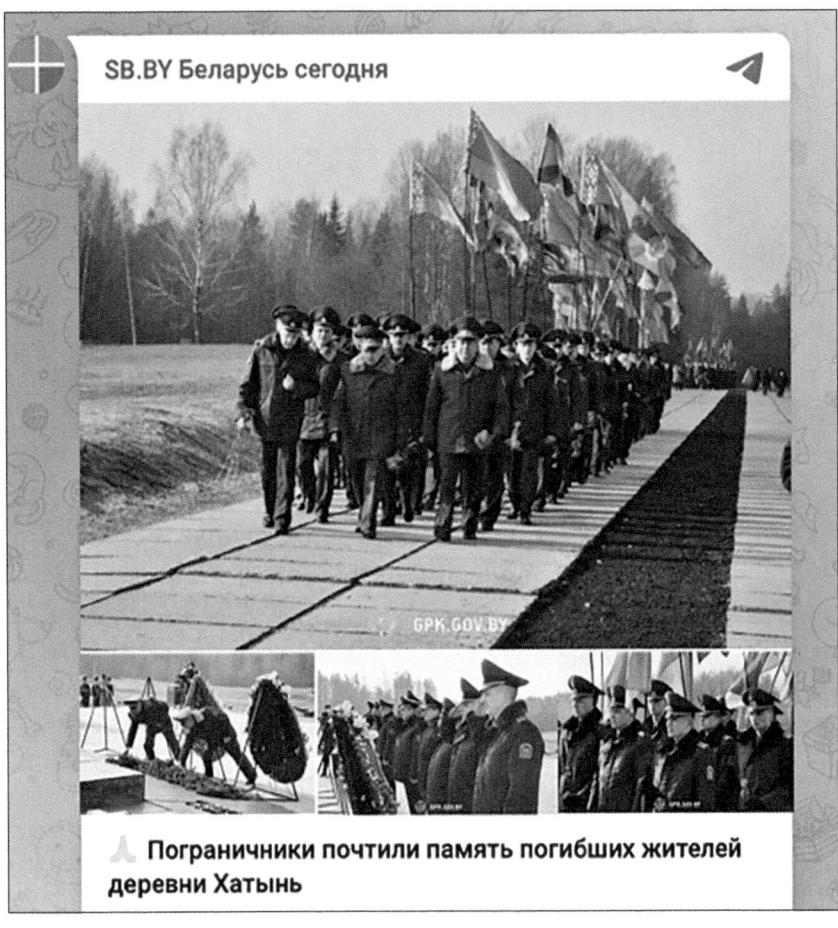

Figure 13: "Belarusian border guards paid tribute to the memory of the perished Khatyn villagers," March 22, 2024. Screenshot. (https://t.me/sbbytoday/112709). Reproduced in accordance with Belarusian copyright provisions.

---

13   Also, the Brest Fortress, another central WWII Memorial was one of the places to visit for international tourists.

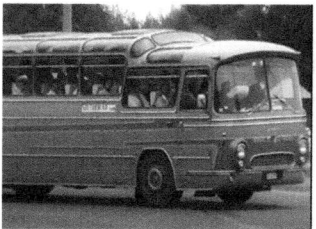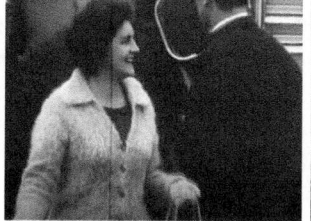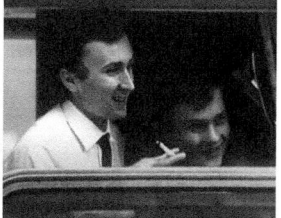

Figure 14: International tourists laugh and smile upon their arrival in Soviet Belarus. Stills from Chatyn, 5th Kilometr (Kolovskii and Adamovich 1968).

The US President Nixon was in good company of Fidel Castro (1972) and Mikhail Gorbachev who visited Khatyn soon after his assession to power. Yet these are ordinary Soviet citizens who were usually pictured against the background of the memorial on the postcards and in various booklets (Fig. 15).

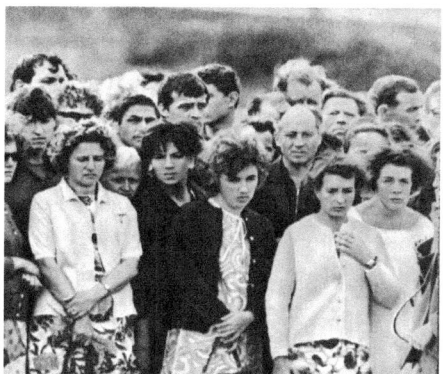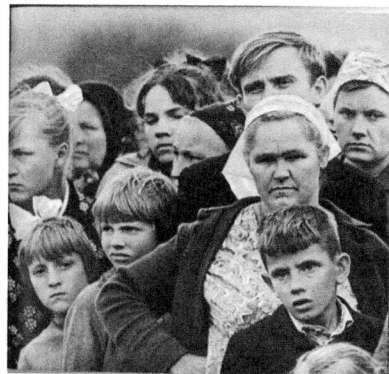

Figure 15: Visitors of Khatyn. Source: Kaziulia Iauhen, Vetsik, Mikhail, Traianouski Ales. Khatyn. Minsk: Belarus, 1977.

In contemporary Belarus the close zooming into the faces of ordinary citizens is rare. Both the Memorial and the people visiting it usually serve as a backdrop for the pro-state officials and journalists to demonstrate their loyalty to the regime. The lay people are then represented by construction workers or create faceless crowds in the distance shadowed by loyal bureaucrats' faces. Similarly, international visitors, be it unnamed "French diplomats", "Russian veterans" or the Gouverneur of the Omsk Oblast Vitalii Khotsenki, are taken to Khatyn to praise the memory politics of the Belarusian state, and personally of its "leader". During the last year, the tendency arose to organize and bring different professional groups from all over Belarus to the Khatyn memorial, including border guards (Fig. 13), university teachers, and bank clerks. Another important target group is schoolchildren and young adults.

The main protagonist of all commemoration events around Khatyn during the last years was and remains the Belarusian incumbent Aliaksandr Lukashenka, who (Fig. 16), demonstrating (together with his extended family) his concerns with the questions of historical justice, clearly embarking on the symbolic value of Khatyn.

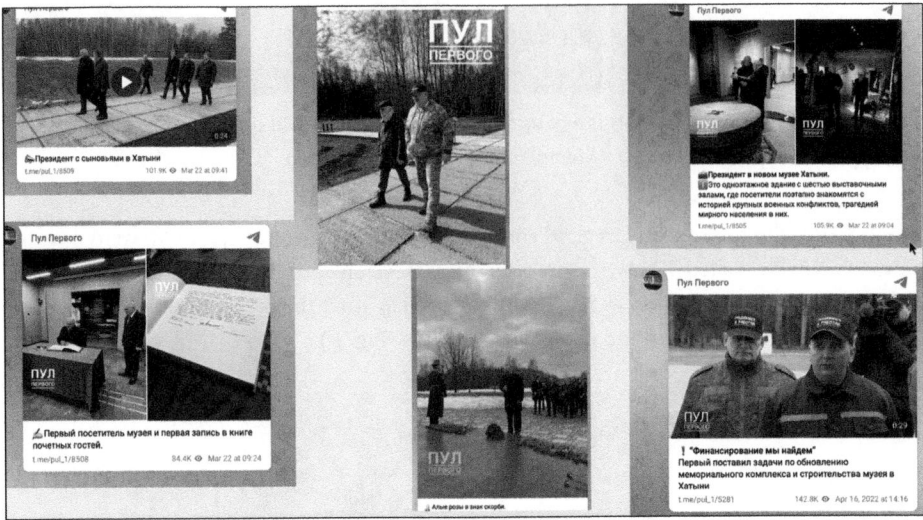

Figure 16: Lukashenka in Khatyn 2022-2023, selected posts' screenshots from @pul_1. Used in accordance with Belarusian copyright provisions.

Lukashenka render to be as a central figure in relation to the Khatyn Memorial – working, building, planting trees, laying flowers, giving orders, inspecting, setting the tone of discussion, and signing the visitor's book. The state-sponsored media make every effort to inscribe him into the history of Khatyn. The most telling example is a recent interactive project launched by the publishing house "Belarus Today" (Belarus segodnia) and timed to the 80th anniversary of the Khatyn Massacre. The high-quality and skillfully visualized digital story-map, symptomatically begins with the "faces of Khatyn killers." Yet a significant part of the story, one of its six main chapters titled "The President in Khatyn" (President v Khatyni) has been fully devoted to Lukashenka's presence in Khatyn. Thus, showcasing the monument of Khatyn, substitutes for history, while, its visual appearance serves as the backdrop for the public performances of Lukashenka.

Conclusion

In their recent article for Government and Opposition, Schlumberger et al. offered a framework for the Study of Digital Dictatorship." (2023). They argue that the digitalization of autocratic regimes means not only that dictatorships are becoming more flexible or more omnipresent, but that their very nature is changing. In this article and on the example of the memory of the executed by Nazis in the village of Khatyn I attempted to demonstrate how Soviet commemoration politics and especially its visual component has been performed and transformed under the new digital conditions.

It is likely that among other things, it is the market value of the Khatyn brand, its recognizability in the post-Soviet space and internationally that attracted the attention of Aliaksandr Lukashenka and his propaganda apparatus. As I have written elsewhere (Astrouskaya 2024) throughout the thirty years of its existence, the Lukashenka regime has failed to design effective cultural politics and was unable to create new recognizable symbols of its own (except perhaps of the figure of Lukashenka himself). Notwithstanding, the authoritarian regime in Belarus continues to embark on visually powerful and it is still appealing to many Soviet brands, refilling them with arbitrary but pragmatic interpretations and accumulating both its Soviet, non-Soviet, and grassroots (off-Soviet) elements. The "genocide" rhetoric, the suffering of the peaceful population has been revoked again and again to spread the fear and anxiety among the population reinforced by the state terror in the aftermath of the peaceful protests of 2020 and the Russian War on the neighboring Ukraine. The traumatic experience of Belarus in the Twentieth century, including two World Wars, Hunger, and Repression makes the Belarusian population highly susceptible to these calls.

Although borrowing, reproducing, and "radical shifting of the general line" (Friedrich and Brzezinski 1956: 134) is commonly characteristic of autocratic and dictatorial regimes, the digital condition brings a fundamental change (Schlumberger et al. 2023: 3), creating a situation when every reference, collation with the original is deemed excessive and every possible rule is disregarded.

Social media platforms such as Telegram endowed autocratic regimes, including that of Lukashenka, with the flexibility and omnipresence they never had earlier. The telegram copy-pasting, as I would like to call it, the reproduction and dissemination of indefinite copies and the producing and sharing of arbitrary interpretations of historical events without paying attention to the original is an essential part of this new visual and knowledge regime. In the case of the Khatyn story already Soviet representations embodied in multiple publications, postcards, and the Memorial were radically detached from the original event. In the case of the Belarusian propaganda channels and despite their seeming attachment to the Soviet past, the Soviet interpretation similarly loses its relevance.

References

Ackermann, Felix. "Der Genozid am Belarusischen Volk" als politischer Diskurs und Strafverfolgungspraxis. In: *Belarus-Analysen* 56 (2021), 2-6.
Adamovich, Ales, Bryl Ianka and Kalesnik Uladzimir. Ia z vohnennai veski. Minsk: Mastatskaia litaratura, 1975.

Arendt, Hannah. Lying in Politics, The New York Review, November 18, 1971, https://www.nybooks.com/articles/1971/11/18/lying-in-politics-reflections-on-the-pentagon-pape/.

Astrouskaya, Tatsiana. Culture as Battlefield. Belarus's Struggle Over Culture and Identity. In: SCEEUS Guest Commentary 12 (2024), https://sceeus.se/en/publications/culture-as-a-battlefield-belaruss-struggle-over-culture-and-identity/.

Astrouskaya, Tatsiana. In Schoolbooks and on Telegram: What Is the Place of Ukraine and Ukrainians in the Memory Politics of Post-Soviet Belarus? In: *Journal of Applied History* 4, 1–2 (2022): 9-27.

Bourdieu, Pierre. On Television. New York: The New Press, 1998.

de Certeau, M. The Practice of Everyday Life. Berkley, Los Angeles&London: University of California Press, 1988.

Chernyshova, Natalya. Between Soviet and Ethnic: Cultural Policies and National Identity Building in Soviet Belarus under Petr Masherau, 1965–80. In: *Kritika: Explorations in Russian and Eurasian History*, vol. 24 no. 3, 2023, p. 545-584.

Edele, Mark. Fighting Russia's History Wars: Vladimir Putin and the Codification of World War II. In: *History and Memory*, vol. 29, no. 2 (2017), 90–124.

Exeler, Franziska. 'What Did You Do during the War?' Personal Responses to the Aftermath of Nazi Occupation. In: *Kritika: Explorations in Russian and Eurasian History* 17, no. 4 (2016), 805–835.

Ford, Matthew and Hoskins, Andrew. Radical War. Data, Attention and Control in the Twenty-First Century. London: Hurst, 2022.

Fickers, Andreas. What the D does to history: Das digitale Zeitalter als neues historisches Zeitregime?. In: *Digital History: Konzepte, Methoden und Kritiken Digitaler Geschichtswissenschaft*, edited by Karoline Dominika Döring, Stefan Haas, Mareike König and Jörg Wettlaufer, Berlin, Boston: De Gruyter Oldenbourg, 2022, pp. 45-64.

Friedrich, Carl and Brzezinski, Zbigniew. Totalitarian Dictatorships and Autocracy, New York, Washington&London: Praeger, 1965.

Gabowitsch, Mischa. Victory Day: The biography of a Soviet holiday. In: *Eurozine*, https://www.eurozine.com/victory-day-the-biography-of-a-soviet-holiday/, May 8, 2020.

Gabowitch, Mischa, ed. Pamiatnik i prazdnik. Etnographia Dnia Pobedy. Sankt-Peterburg: Nestor-Istoria, 2020.

Gironi, Camilla. A New Belarusian Partisan Republic? World War II Narratives between Myth, Revisionism, and State Propaganda in Lukašenka's Speeches in the Post-2020 Presidential Election Era In: *Acta Albaruthenica* (2023), 123–144.

Goujon, Alexandra. Memorial Narratives of WWII Partisans and Genocide in Belarus. In: *East European Politics and Societies*, 24(1) (2010), 6-25.

Kazharski, Aliaksei. 2023. "An Authoritarian Spectacle: Visual Biopolitics and the Dramaturgy of the Poland-Belarus Border Migration Crisis." *Visual Anthropology* 36 (4): 373–96

Klymenko, Lina. Narrating the Second World War. History Textbooks and Nation Building in Belarus, Russia, and Ukraine. In: *Journal of Educational Media, Memory, and Society*, 6 (2), 2016, 36–57.

Kolovskii, I. (regi), Adamovich A. et al (script). Chatyn, 5-yi kilometr (A Documentary). Belarusfilim: 1968, https://www.youtube.com/watch?v=5RqNu8mqVhw.

Koran, Gleb. Telegram in Belarusian Protests of 2020: Affective Tool for Populist's Uprisings. In: *Digital Icons* 22 (2023): 65-81.

Kragh, Martin. Russia, Katyn and Bucha: How the Kremlin is Falsifying History. https://sceeus.se/en/publications/russia-katyn-and-bucha-how-the-kremlin-is-falsifying-history/ (April 22, 2024).

Kurilla, Ivan. Reusing Soviet History Books: The Role of World War II in Russian Domestic Politics and Academia. In: *The Journal of Slavic Military Studies*, 33(4) (2020), 502–507.

Laputska, Veronica. World War II Criminals in Belarusian Internet Mass-Media: The Cases of Anthony Sawoniuk and Vladimir Katriuk. The Journal of Belarusian Studies 8 (2016), 50-77.

Lewis, Simon. Belarus – Alternative Visions: Nation, Memory and Cosmopolitanism. New York and London: Routledge, 2019.

Lewis, Simon. Overcoming Hegemonic Martyrdom: the Afterlife of Khatyn in Belarusian Memory. In: *Journal of Soviet and Post-Soviet Politics and Society*, 1:2 (2015), 367–401.

Lüer, Jörg. Katyn und Chatyn – Fragen an die gesellschaftliche Bedeutung von Erinnerung. In: *Ost-West europäische Perspektiven* 1 (2002), https://www.owep.de/ausgabe/2002-1.

Makarychev, Andrey. "Visual Biopolitics: Outlining a Research (Sub)field." *Journal of Illiberalism Studies* 1 (2021): 51-57.

Markham, Tim. Social media, protest cultures and political subjectivities of the Arab Spring. *Media, Culture & Society*, 36(1) (2014), 89-104.

Marples, David. "Our Glorious Past." Lukashenka's Belarus and the Great Patriotic War. Stuttgart: ibidem, 2014.

"Muzei, chasovnia, novaia podsvedka – kak izmenilas' Khatyn' posle rekonstruktsii", March 19, 2023 (https://sputnik.by/20230319/muzey-chasovnya-novaya-podsvetka--kak-izmenilas-khatyn-posle-rekonstruktsii-1073329332.html).

Oushakine, Serguei Alex. Postcolonial Enstrangements: Claiming a Space between Stalin and Hitler. In: Buckler, Julie, Johnson Emily D. (eds.) *Rites of Place: Public Commemoration in Russia and Eastern Europe*, Evanston: Northwestern University, 2013, 285–3015.

Ravveduto, Marcello. Past and Present in Digital Public History. In: *Handbook of Digital Public History*, edited by Serge Noiret, Mark Tebeau and Gerben Zaagsma, 131-138. Berlin, Boston: De Gruyter Oldenbourg, 2022.

Rudling, Per Anders. For a Heroic Belarus!: The Great Patriotic War as Identity Marker in the Lukashenka and Soviet Belarusian Discourses. In: Sprawy Narodowościowe 32 (2008), 43–62.

Rudling, Per Anders. The Khatyn Massacre in Belorussia: A Historical Controversy Revisited. In: *Holocaust and Genocide Studies* 26 (1) (2012), 29–58.

Schlumberger, Oliver, Edel, Mirijam, Maati, Ahmed, Saglam, Koray. How Authoritarianism Transforms: A Framework for the Study of Digital Dictatorship. *Government and Opposition*. Published online 2023:1-23. doi:10.1017/gov.2023.20.

Soldatov, Andrei, Borogan Irina. The Red Web. The Kremlin's War on the Internet. New York: Political Affairs, 2015.

Wijermars, Mariëlle, and Tetyana Lokot. 2022. Is Telegram a 'Harbinger of Freedom'? The Performance, Practices, and Perception of Platforms as Political Actors in Authoritarian States. In: *Post-Soviet Affairs* 38 (1–2): 125–45.

Wolfsfeld, Gadi, Segev, Elad, & Sheafer, Tamin. (2013). Social Media and the Arab Spring: Politics Comes First. *The International Journal of Press/Politics*, 18(2) (2013), 115-137.

Weiss-Wendt, Anton. *Putin's Russia and the Falsification of History: Reasserting Control over the Past*. London: Bloomsbury Academic Press, 2020.

"80 let boli. Projekt izdatel'skogo doma 'Belarus segondnia'", http://sp.sb.by/khatyn80#6 (n.d.) [2023?]

Izolyatsia: A Factory, But a Different One
Production of Images by the Means of Violence

Lesia Kulchynska

Abstract

The paper explores the case of Izolyatsia, a Soviet-time insulation materials factory in the city of Donetsk, which was transformed into an art center in 2010, and later, in 2014, was seized by the representatives of the Donetsk People's Republic [1] (DPR) and turned into a torture prison controlled by the State Security Ministry of the DPR supervised by the Russian Federal Security Service. The author traces the transition of Izolyatsia from industrial production to immaterial production of lifestyles and meaning to the production of violence and, at the same time, the production of images through violence. Referring to the evidence of the former prisoners of Izolyatsia, the author shows how captives of Izolyatsia are being used as source material for the propagandistic images, and how tortures are used as a means to extract those images. The author introduces the concept of „image extraction infrastructure" to analyze the conversion of practices of violence and image production within the operation of the Izolyatsia torture camp, showing how the torture rooms of Izolyatsia, but also across the Russian-occupied territories, function as an extension of the Russian state-controlled media complex as image-production facilities.

Keyword

visuality of violence, media, Donetsk People's Republic, image extraction, immaterial production

1.

In her famous essay "Is Museum a Factory?", published back in 2009, Hito Steyerl commented on a worldwide trend of transforming abandoned factory premises into museums of contemporary art as a symptom of a broader transition from material to immaterial production. She wrote that the museum became a 'hotbed of contemporary production. Of *jargon, lifestyles, and values*"[1]; "a factory, so to speak, but a different one[2]". It was an apt observation, indeed, which grasped the emerging pervasiveness of the idea of art as a (presumably) effective instrument of shaping, changing, transforming, and altogether fabricating social reality through the operations with images, words, senses, visions, and bodies.

In 2010 art center *Izolyatsia*(eng: isolation, insulation) was created in the Ukrainian city of Donetsk on the site of the former insulating materials factory, perfectly fitting the described tendency. The founder of the art center, Lubov Mykhailova describes the city of Donetsk as a postindustrial space, meaning that the entire infrastructure of the city was built during the soviet times around the plants and factories, most of which ceased to function properly in the 1990-is. Once those industries collapsed, the people were left abandoned and puzzled; "the state never offered them any alternative[3]", said Lubov. *Izolyatsia* came to fill this gap and create an alternative to the outdated industrial way of life. "We strongly believe that culture is a very powerful tool for social change"[4], said Lubov and this belief was a flagship principle of the art center activity.

In the documentary film about the *Izolyatsia* the most repeated words, – just like in many contemporary art press releases – are *"alternative"*, *"change"* and *"transformation"*[5]. The artists and other creative workers who collaborated with the art center in concert describe its activity as an endeavor to transform the life of the city and its inhabitants through art.

Designer Maryna Samokhina says that *Izolyatsia* became a "place that offered an alternative way of seeing the future[6]". "Even in the very practical sense, as a personal choice of the occupation. It turned out it's possible not to go to work to the plant, but do something else"[7].

When *Izolyatsia* had to stop its activity in Donetsk for the reason we will discuss soon, its team initiated the project "Change" aimed at fostering changes, namely a transition from an industrial to a creative economy, in other eastern postindustrial cities.

To put it shortly, referring to Hito Steyerl's metaphor, the *Izolyatsia* art center, located at the former insulating materials factory, remained a "factory, so to speak,

---

1 Hity Steyerl, "Is Museum a Factory?", e-flux Journal, #7 (2009), https://www.e-flux.com/journal/07/61390/is-a-museum-a-factory/
2 Ibid.
3 Diana Kuryshko. "The founder of "Isolation": we were strangers to "DNR"" (Interview with Lubov Mykhailova), BBC, June 9, 2017, https://www.bbc.com/ukrainian/features-40227810
4 Ibid.
5 Sergiy Ivanov, "Concentration Camp "Izolyatsia"", documentary, 2010, Концтабір "ІЗОЛЯЦІЯ" | Документальний фільм
6 Ibid. 9:39
7 Ibid. 9:39 - 9:46

but a different one". Instead of producing insulating materials, it transitioned to the production of new, alternative, subjectivities, visions, economies, and ways of living. At the same time, it was meant to become a point from which the transition of the entire region from industrial to creative economy, or from material to immaterial production, could have sprung.

2.

Yet, the story went in quite a different direction. Almost immediately after the beginning of the Russian invasion of Donbas, the site of the *Izolyatsia* art center was seized by the representatives of the Donetsk People's Republic (DPR) and turned into a torture prison controlled by the State Security Ministry of the DPR and supervised by the Russian Federal Security Service. Many of its captives described it as a concentration camp based on the fact that detention there doesn't imply any lawful procedure, and its operation is not regulated by any legal documents. You can be placed there without any legal conviction and even explanation, and are not granted any rights, such as the right to call your relatives or the right to call for an advocate.

The seizure of the art center happened on June 9, 2014. That day a group of armed men entered the territory of the institution. "They showed a piece of paper that said: eight factory hectares now belong to the DPR"[8], recalls Mykhaili Gluboky, head of the development and communications of the art center.

Leonid Baranov, a head of the so-called, Special Committee, created, as he explained, to fight "all the saboteurs and traitors inside and outside of the system"[9], which quartered in Izolyatsia, commented on the seizure in his interview with the Russian Forbes: "This is the same plant about which they say that it was almost the center of the world art, which we allegedly occupied. Considering what kind of art it was, it was impossible not to occupy it"[10]. The interview is recorded on video, and we can see how he takes the catalog of Eva and Adele's exhibition named "Futuring" as an example of this "kind of art" and says: "In our republic, we can live without it"[11]. Therefore, the transformation of the art center into a torture prison started with the destruction of the artworks.

---

8   Ukrainska Pravda, "Fear and death in "Isolation". How people are tortured in the basements of Donetsk", Ukrainska Pravda, February 24, 2020, https://www.pravda.com.ua/rus/articles/2020/02/24/7241046/
9   Izolyatsia, "Leonid Baranov at Izolyatsia", https://vimeo.com/102014455
10 Ibid.
11 Ibid.

3.

Among the first spectacularly destructive actions of the new administration of the former art center was a detonation of the site-specific sculpture by the Cameroonian artist Pascale Marthine Tayou with a symbolic title in this context *"Make Up... Peace"*. The piece was a giant red lipstick placed on top of the no longer functioning boiler house chimney. The artist said it was his tribute to the women of Donetsk who played a major role in the rebuilding of the city after World War Two[12]. This gesture can be read as an ironic transformation of the patriarchal symbol of phallic domination into a manifestation of women's emancipation. It was also an apparent sign of the deindustrialization of the place and its shift from the industrial economy to the creative one, or from the industrial production to the production of images.

"From my point of view, Donetsk is not just a city of mines and scrap metal, but also an island of dreams"[13], said Pascale Marthine Tayou. Against the backdrop of the many years of Soviet industrial exploitation, accompanied by the official ethics of labor self-sacrifice and asceticism, the shiny red lipstick probably evoked dreams more related to leisure and consumption, promised to be fulfilled by IZOLYATSIA as a site of creative pastime. Although, the consumption of artistic content is another type of labor, as Hito Steyerl argues, for those who worked at the Donetsk plants, the difference between the factory and the art center was yet too drastic. The factory chimney transformed into lipstick marked this difference, as well as the role Izolyatsia Art Center played in the life of Donetsk and its postindustrial transformations. It's claimed on the art center website that 'Lipstick' "became a topic of discussion, as well as a Donetsk landmark, which could be seen from anywhere in the city", and "one of the most recognizable symbols of the 'Izolyatsia[14].

In 2015 the "Lipstick" was blown up by the military invaders of the pace, clearly signaling that the time for dreams was over. The site of leisure turned back into a place of labor, this time a coerced one. The spectacular moment of the explosion was filmed on the phone by its authors and uploaded to the Internet. Of course, the video created by the terrorists became viral, capitalizing on the existing interest in the well-known art center and a widely discussed sculpture.

If the artist Pascale Marthine Tayou attempted to redefine the meaning of the existing object through material intervention, the Izolyatsia invaders went even further, displacing the material object altogether with a digital image. Both the chimney and the lipstick were all gone and have been replaced by the free-floating image of the violent destruction of what was once a "Donetsk landmark". The

---

12 https://izolyatsia.ui.org.ua/ua/lipstick/#prison
13 ibid.
14 Ibid.

feminine symbol of Izoliatsia's transformation was replaced by the media image of domination of the well-armed group of men.

As another manifestation of violent patriarchal dominance, the invaders used the sculptures from Maria Kulikowska's project Homo Bulla as shooting targets. The sculptures were a cast of the artist's body made of soap, a fragile material symbolizing human vulnerability and mortality. One of the militants, who calls himself Mongol told the Russian Forbes journalist while shooting the sculptures that it was his performance. When Mongol was asked by the journalist what he thinks about Izolyatsia, he answered: "The art of European integrated, advanced Maidan people? Let them practice this art on their Maidan until we get there"[15]. Apparently, his performance as well as the blow-up of the *"Make Up... Peace"* sculpture was a sort of performative message to the "Maidan people" in Donetsk and beyond.

4.

Although the present-day Izolyatsia holds the status of a *secret* prison, its operation was accompanied by an extensive media presence from the very beginning. Almost immediately after the seizure, its new administration was interviewed by the Russian Forbes, commenting on the changes in the site's functioning. It was important for invaders to make a public statement explaining their occupation of the art center. The main argument publicly announced was not the military needs, but the troublesome activity of the art center and art exhibited there.

To prove the rightfulness of their invasion, Baranov chose to focus on Boris Mikhailov's book "Look at Me I Look at Water. Or Perversion of Repose" found by him in Izolyatsia. Flipping the pages of the book in front of the camera, he says: *"For me, this is not art, and cannot be art, this is sick people shooting and showing to other sick people. On the territory of DNR drug abuse and this kind of art will be prohibited because this is not art, this is pornography. Therefore, we could do anything else but drive those sick-headed people away from here"*[16].

While Baranov is saying this, the camera scrutinizes the images from the book: an image of a naked aging body of Boris Mykhailov posing; and a naked woman holding a watermelon; special attention is given by Baranov to the picture of a woman peeing in a bucket.

Stanislav Asyeev, a former prisoner of Izolyatsia, describes in his book *"The Torture Camp On Paradise Street"* what kind of activity replaced the expelled art: "Here's another example. A man had been tortured for several hours and had signed everything that was placed before him. And yet, this wasn't enough. After the torture, the guards carted the guy off into a cell, stripped him naked, put

---

15 Video from Izolyatsia's archive
16 https://izolyatsia.ui.org.ua/ua/lipstick/#prison

on some music on a cell phone, and made him dance for the camera"[17]. Another former prisoner, Valentyna Buchok, recalls her experience: "There was no toilet in the prison cell, only a 5-liter bucket; and a 2-liter bottle; you pee in the small one and then pour it in the big one; and there is a camera in the corner, the light is always on"[18].

Notably, all the violence that takes place in Izolyatsia is inextricably linked with the practices of image production. The images of art prohibited in DNR were replaced with images of violence produced by DNR's Russian-backed administration.

5.

Borys Mykhailov's book was not by chance chosen by Baranov as evidence of Izolatsia's hostile intentions. Being the most famous and acknowledged post-soviet contemporary art photographer in the West, Mykhailov is often claimed to have gained the interest of Western viewers by exposing the collapse of the Soviet project and uncovering the shadow side of the "socialist utopia". He was many times criticized for the "exploitation" of the post-soviet misery and exoticizing it for the "Western gaze".

Back in 1995, Mykhailov's exhibition "Me Not Me" in the Kharkiv Museum of Art was banned by the then-director of the museum Valentima Myzgina. She explained the "unworthy" content of the show as an implementation of the so-called "Dulles plan', pointing out that he proposed this show to the Museum right after his trip to the USA[19]. *Dulles's plan* is a fictional document of a conspiracy theory, according to which the CIA chief Allen Dulles had developed a plan for the United States to destroy the Soviet Union during the Cold War by secretly corrupting the cultural heritage and *moral values* of the Soviet nation.

Obviously, there was no such thing as the Soviet Union in 1995 anymore. The so-called "moral values", Mrs. Myzgina was worried about referred rather to the specific visual regime traced back to Soviet times, which was threatened by the scandalous arrival of contemporary art (from the West). While the Soviet project was based and totally dependent on the operation of repression, both in political and psychoanalytical senses, excluding from the field of vision everything that contradicted an officially promoted fantasy, contemporary art, on the contrary, brought the gaze particularly interested in what was repressed and excluded from the official picture. Borys Mykhailov perfectly adopted that kind of gaze to look

---

17   Stanislav Aseyev, *The Torture Camp On Paradise Street*, Ukrainian Research Institute Harvard University, 2023, p. 185

18   Sergiy Ivanov, "Concentration Camp "Izolyatsia"", documentary, 2010, Концтабір "ІЗОЛЯЦІЯ" | Документальний фільм

19   Video Inetrview from Lesia Kulchynska'a archive

at his native environment. While the post-soviet beholders, accustomed to the socialist realism visuality, expected from art a made-up image of reality covering its problematic parts, Borys Mykhailov, on the contrary, offered an image exposing the most uncomfortable sides of the post-soviet reality and its inhabitants. His success in the West made his art even more painful for those who recognized themselves in his images: the pain of being seen without cover was enhanced by the pain of being exposed in all their misery to the supposedly dismissive Western sight.

A helpful way to cope with this kind of "hostile gaze" is to attribute it to the "enemy", as both Myzgina and Baranov did. *"Million-dollar grants have been involved here; whoever possible and not possible conducted their training sessions here. Those people hate everything Slavic, everything Russian. Therefore, we could not watch all this and stand aside,"*[20] said Leonid Baranov about Izolyatsia while showing the journalists Mykhailov's book as proof of his words.

6.

While Mykhailov, as a bearer of a critical gaze, epitomizes for the self-proclaimed authorities the figure of the enemy, it is no surprise that espionage is the most common sentence for their prisoners. This sentence can be attributed without any legal procedure to everyone who doesn't openly support DNR or came there without special permission from other parts of Ukraine or other countries. In other words, the very act of seeing could be deemed as espionage if you are not part of DNR power structures.

The former prisoner of Izolyatsia, Stanislav Aseyev described his experience of being a prisoner at Iaolyatisa in his book "Tortures Camp on Paradise Street". He wrote that the main rule of the prison is a strict prohibition to observe: "In Izolyatsia, we were ordered not to look at the windows (never mind that they were painted over), or at the CCTV camera, or at the tray slot— the small opening in the door through which the guards delivered food. Because each of these objects was located on a different wall, we were compelled to stare straight ahead in space or at the floor, silently and without moving"[21]. Despite the fact that the administration staff of Izolyarisa wears balaclavas covering their faces, during the encounters with them, the prisoners always have to have plastic bags on their heads. Those bags are handed to them immediately upon arrival and should be carefully preserved. "As soon as a door opens, everyone in the cell jumps to their feet, pulls the bags over their heads, places their hands behind their backs, and turns to face

---

20   Izolyatsia, "Leonid Baranov at Izolyatsia", https://vimeo.com/102014455
21   Stanislav Aseyev, *The Torture Camp On Paradise Street*, Ukrainian Research Institute Harvard University, 2023, p. 36

the window"[22], writes Aseyev. The self-proclaimed authorities reserve the right to look for themselves, while those who are conquered should remain at the position of the object, exposed and accessible for the gaze but never returning it.

The same visual dispositive is established on all Russian-occupied territories in Ukraine. My friend who visited Kherson after de-occupation wrote on her Instagram: "In the eyes of everyone I met was something that is quite hard to describe. The traces of fear these people lived with for many months... People were talking about hiding their phones in their underwear while going out. About avoiding looking at the Russians – you could be taken only for that.[23]"

Russian soldier named Sergey, who took part in the occupation of Bucha told his girlfriend in the phone conversation intercepted by Ukrainian security that his command gave them an order to kill every civilian they saw: "They told us that, where we're going, there's a lot of civilians walking around. And they gave us the order to kill everyone we see". "Why the fuck?", asked his girlfriend. "Because they might give away our positions"[24]. To be killed you don't even have to look at the invaders, it is enough just to have eyes and be able to see. The very capacity to look is already threatening to the invaders to such an extent that the one who possesses it should be eliminated. Under Russian occupation, you can maintain your existence only with a trash bag on your head.

7.

In Izolyatsia the prohibition to look is coupled with the permanent and all-encompassind surveillance. While the prosecutors are hidden from the eyes of the prisoners, the later are turned into an everlasting spectacle.

Recalling his experience of being a captive at the Izolyatsia concentration camp, Stanislav Aseyev admits that the most mind-blowing thing for him was to realize that all the terror that was going on there daily was being filmed: "Sometimes we felt like we were in the middle of an experiment; the utter unreality of what was happening and being filmed with a dozen cameras would make us believe that"[25].

Describing the most terrifying tortures, he emphasizes: "And all of it is documented, all of it captured— there's a camera in each cell, in each solitary, in every basement. *There are terabytes of recordings and hundreds of hours for international*

---

22 Ibid. p. 4

23 @ooleksandra_kravchenko, Instagram, November 20, 2020, https://www.instagram.com/p/ClMy6eNtb_w/?utm_source=ig_web_copy_link&igsh=MzRlODBiNWFlZA==

24 Yousur Al-Hlou, Masha Froliak and Evan Hill. *'Putin Is a Fool': Intercepted Calls Reveal Russian Army in Disarray*, New York Times, Sep. 28, 2022, https://www.nytimes.com/interactive/2022/09/28/world/europe/russian-soldiers-phone-calls-ukraine.html

25 Aseyev, p.XVII

courts. This, too, feels like an experiment: Can these people really go on filming their crimes, with utter impunity, for six years, laughing off every UN report? It would appear that they can, in which case Isolation tells you what our world really is. All its meaninglessness, its cruelty, and injustice are concentrated right here, at 3 Paradise Street. There is no retribution, only the mockery of us, the defeated.[26]"

The purpose of the extensive video archives of violence produced at the Izolyatisa torture camp is still to be expored. We can suggest, that their destiny will be defined by the outcome of the war. Will those images serve as a testimony of crimes of the perpetrators or will they be used as a visual proof of their power depends on whether the Russia's terror will bring it victory.

Meanwhile, when the war is still ongoing, the cameras in the prison cells function as tools of surveillance as well as appropriation. Even if the one got released from imprisonment or simply was killed there, his or her image remains the property of the perpetrators. Being the evidence of violence those images are also testimonies of humiliation of the imprisoned. Those images can be used for blackmail and intimidation, serving as a tool of power.

"It didn't matter who or what you were before coming here[27]", wrote Aseyev. In the weird movie filmed by the prison's cameras you are captured in the role of a prisoner, a defeated enemy turned into a convicted criminal. By transforming individuals into characters of a torture horror documentary, the self-proclaimed authorities classify the reality in their own manner and visualize the order established by them by means of pure violence.

8.

The trashbags on the heads of the imprisoned, as well as balaclavas covering the faces of prosecutors, signal that they recognize the possibility of legal prosecution and are trying to take protective measures. Indeed, at the moment the main organizer of the torture operations, nicknamed Palysh, is being sentenced by a Ukrainian court to 15 years of detention.

Yet the prison once run by him continues to operate. Despite its illegal status, its operation is followed by extensive media coverage. The managers of the site, as well as former prisoners, give interviews to the established media. The journalists being capt there are often released and actively spreading first-hand evidence about all the human rights violations taking place there. Dmytro Potiekhin, a journalist and another ex-prisoner says that while being released, he was specifically asked by his prosecutor to tell the truth about what he saw and experienced there, which means that the violence taking place there is not meant to be hidden. Quite on the contrary. Aseyev recalls that one of the Izolyatsia guards was bragging that

---

26 Ibid.
27 Ibid.p. 29

"even bus drivers avoided stopping anywhere near it. The terrible reputation of the "Donetsk Dachau" makes its founders truly proud". "The terrible reputation" is another immaterial product being manufactured in this peculiar factory on purpose. It is designed to induce fear, which is a main source of the oppressor's power.

9.

Describing the operation of the Izolyatsia concentration camp Aseyev points out the outstanding randomness of its captives, which created an impression that literally anyone, regardless of his or her activity or political views could end up there by some bad luck. Pro-Ukrainian views or registration outside DNR would definitely increase your chances; yet, he claims that many of those newly arrived were genuinely confused about why they've been detained and believed it was some kind of a mistake. The option to appeal the decision or call an advocate is not available there.

What seemed to him even more striking was the randomness of torture victim selection: "Torture was inflicted on military servicemen, long-haul drivers, business owners, and doctors. It was inflicted on everyone, for no reason whatsoever". "This evident chaos has one specific goal: to make you lie"[28], comments Aseyev.

Valentyna Buchok, a former prisoner recalls that during the torture session, she was directly given the advice to make up the confession about her non-existent collaboration with the Ukrainian security service to stop the torture. This made-up confession needed to be recorded on video and documented in the signed protocol.

An ultimate goal of the Izolyasia's prosecutors is, therefore, not the extraction of some strategic information, not even fabrication of loyalty through fear. The value extracted from the imprisoned human beings by means of torture is an image designed to support the Russian propaganda version of reality.

"Torture rooms, for example, operate simultaneously as machines for extracting information from people, and as the mises-en-scène for Russian propaganda TV, which broadcasts information placed forcibly into the mouths and bodies of disposable war subjects. These subjects are compelled to articulate messages or confessions by means of electric current, water torture, rape, hunger, broken bones, and cut flesh"[29], writes Svitlana Matvienko. Torture rooms across the Russian-occupied territories function as an extension of the Russian state-controlled media complex. Izolyatsia is a part of its image-extraction infrastructure.

---

28 Ibid., p. 104
29 Svitlana Matviyenko, *Speeds and Vectors of Energy Terrorism*, e-flux, #134, March 2023, https://www.e-flux.com/journal/134/525421/speeds-and-vectors-of-energy-terrorism/

According to Aseyev, Izolyatsia is a peculiar concentration camp because its operation is not based on any differential principle. Its captives don't share either national, religious, or even political identity. It is a random bunch of people, imprisoned there for different reasons and turned into source material for images of Russian "enemies". Violence, in this case, becomes an instrument of image extraction, a tool for processing humans into images needed for the war machine to operate.

10.

Writing about visuality as an instrument of authority, Mirzoeff claims that it operates through classification, naming, and categorizing. Commenting on the functioning of the Russian filtration camps, Svitlana Matvienko and Daria Genmanova point to the operation of labeling of the population as its main activity: "What if, indeed, the purpose of the massive filtration machine is merely to produce *two inexistent groups* of people ... by *falsifying linkages between the labels propagated by Russian state* media or governmental officials and *people's physical bodies*"[30]. The two groups of people they mentioned are: "Nazis" and their "victims" ("the Russian-speaking Ukrainians whom the Russian Federation allegedly came to protect"[31]). During the interrogations in the filtration camps those subjected to filtration are offered two main options to choose their further destiny: those who are identified as "victims" got the permission to stay at the territory of RF, while those identified as "nazis" (mostly, activists, former and current military and security, journalists, or governmental officials) have to undergo the "so-called "denazification courses," after which they also pass "loyalty tests," that are often video-recorded and publicized"[32].

The *linkages between the labels propagated by Russian state* media or governmental officials and *people's physical bodies* are established through the signed documents and images of video-recorded confessions. The latter are needed to make the nonexistent groups of people visible. Within this infrastructure, Izolyatsia's main specialization is the production of images of "Nazis". Yet, it is important to note that it is not only "people's *physical bodies*" that are being used as material for the propaganda image, as Matvienko and Getmanova claim, but entire personalities, with their social profiles, biographies, legal statuses, etc.

Here we face a new modality of image production, which exceeds the familiar logic of fake. It is no more the case of the falsified representation. The people

---

30 Getmanova, D., & Matviyenko, S. (2022). Producing the Subject of Deportation. Filtration Processes during the Russia-Ukraine War. *Sociologica*, 16(2), 239–252. https://doi.org/10.6092/issn.1971-8853/15387
31 Ibid.
32 Ibid.

in the confession videos represent themselves, not someone else; it is not their representations but their personas are being altered. What is being manipulated is not the image, but the very source of it, – the real human being. The operation of producing the needed image relies here on the remaking of the reality itself with the means of violence.

11.

The war is a lawmaking endeavor. It violates the existing law to establish a new one. Within the Russian war on Ukraine, this process of lawmaking relies on the global communication network, promoting the convergence of violence and image: while images serve as the fuel and the license for the industry of violence, violence becomes an integral part of the image production cycle. Within this vicious cycle, Izolyatsia, transformed from the art center into a torture prison, paradoxically remains a site of an immaterial "content production", with its activity grounded in the strong belief in the power of image as a tool for change.

The images of the enemies, fabricated at Izolyatsia by the means of torture are operational as a visual legitimation of a real, well-documented violence, taking place in Izolyatsia cells, as well as on the territory of Ukraine in general. Once the enemy is labeled and visualized, and its image is linked to real human beings the *"terabytes of recordings"* of documented violence, mentioned by the Izolyatia prisoner, can transform from the evidence of a crime to the the evidence of justice, self-defense, population protection, and, of course, power. Although, at the same time, the images and messages of violence, produced and spread by the Izolyatsia concentration camp, function to announce the arrival of power which is not relaying on legitimacy, but on the pure terror.

References

Aseyev, Stanislav (2023): The Torture Camp On Paradise Street, Ukrainian Research Institute Harvard University.

Getmanova, Matviyenko (2022): Producing the Subject of Deportation. Filtration Processes during the Russia-Ukraine War . *Sociologica*, 16(2), 239–252 (https://doi.org/10.6092/issn.1971-8853/15387)

Ivanov, Sergiy (2010): "Concentration Camp "Izolyatsia"", documentary, (https://www.youtube.com/watch?v=bRv_Ui_A4VQ)

"Leonid Baranov at Izolyatsia", (https://vimeo.com/102014455).

"Make up ... Peace", https://izolyatsia.ui.org.ua/ua/lipstick/#prison

Matviyenko, Svitlana (2023): "Speeds and Vectors of Energy Terrorism", e-flux, #134, March 2023, (https://www.e-flux.com/journal/134/525421/speeds-and-vectors-of-energy-terrorism/).

@ooleksandra_kravchenko, Instagram, November 20, 2022, (https://www.instagram.com/p/ClMy6eNtb_w/?utm_source=ig_web_copy_link&igsh=MzRlODBiNWFlZA==)

"'Putin Is a Fool': Intercepted Calls Reveal Russian Army in Disarray", New York Times, Sep. 28, 2022, (https://www.nytimes.com/interactive/2022/09/28/world/europe/russian-soldiers-phone-calls-ukraine.html).

Steyerl, Hito (2009): "Is Museum a Factory?", e-flux Journal, #7, 2009, (https://www.e-flux.com/journal/07/61390/is-a-museum-a-factory).

Засновниця «Ізоляції»: ми були чужими для «ДНР» BBC, June 9, 2017, (https://www.bbc.com/ukrainian/features-40227810).

"Страх и смерть в «Изоляции». Как пытают людей в подвалах Донецка, February 24, 2020, (https://www.pravda.com.ua/rus/articles/2020/02/24/7241046/).

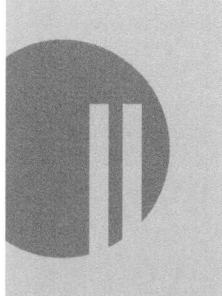

Discourse Analysis
and Social Media

Sensitive Subjects
Talking about the Residents of the Territories Occupied by Russia in Official Ukrainian Discourse

Yana Lysenko

Abstract

This study explores the representation and perceptions of the residents of territories occupied and annexed by Russia since 2014 and after 2022 in the official Ukrainian discourse since the large-scale Russian invasion. Focusing on narratives related to the occupied regions, it examines if and how Ukrainian official communication addresses the very sensitive subjects – the future reintegration measures concerning this population group. By analysing statements from key formal and informal political actors within President Zelensky's government, the study employs an extended framing analysis to capture the main characteristics of the official Ukrainian discourse on the residents of occupied territories from February 2022 to December 2023. The research reveals a dominance of moral judgements and proposed military solutions over discussions on the causes of occupation or detailed plans for reintegration in the studied discourse. Findings indicate a focus on reclaiming territory, with specific attention to Crimea, and a general underrepresentation of the occupied territories' residents in the discourse. The study highlights a lack of a clear action strategy for addressing the reintegration of these residents and suggests a need for more comprehensive research on political discourse concerning the residents of the occupied territories, including broader perspectives and the reasons for their underrepresentation in official narratives.

Keywords

residents of the occupied territories; Russian Invasion; official Ukrainian discourse; reintegration

Introduction

Since the beginning of the large-scale Russian invasion in 2022, the narrative of social cohesion has dominated public discourse in Ukraine. While the intentions to retake the Ukrainian territories that have been occupied since 2014, as well as those claimed after 2022, are openly and clearly communicated at all political levels of Ukraine, the question of how to deal with the inhabitants of these territories has faded into the background of official Ukrainian political communication. Since the imposition of martial law in Ukraine, most political decision-making power has been concentrated in the hands of Ukrainian President Volodymyr Zelensky and his government. Therefore, the current study focuses on the public statements of formal and informal political actors who, as a part of the executive branch under Zelensky, have most shaped the Ukrainian public discourse on political issues since February 24, 2022.

The aim of this study is to examine the representation and perceptions of the residents of territories occupied and annexed by Russia in current official discourse from the Ukrainian government. This study focuses specifically on the narratives relating to the residents of all Ukrainian territories occupied and annexed by Russia since 2014 and after 2022.

This study seeks to explore how and to what extent the selected actors talk about the residents of the occupied territories and whether their statements represent a coherent discourse from which clear communication and an action strategy of Ukrainian government regarding this population group can be derived. This means that on the one hand, the study aims to capture problems and measures discussed in relation to the population targeted by the discourse. Of particular interest is the question of whether there is a clear vision within the discourse for the future reintegration of the residents of the occupied territories and what this might look like. Regarding this question, it is also assumed that dealing with collaborators represents one of the most sensitive issues, suggesting that these topics might be avoided or scarcely addressed in political discourse due to their potential for creating tensions within Ukrainian society.

On the other hand, this study intends to analyse what characterizes the current political discourse on this topic and which frames shape it. It aims to examine who is addressed in the selected statements, how their content is structured and what intentions might lie behind these statements. To explore these research questions, written statements from the chosen actors were selected as a sample. These statements were published on official websites and official Telegram and Facebook accounts, spanning from February 24, 2022, to December 31, 2023. Additionally, when compiling the sample, search words were also defined in reference to the target territories and issues, which were used to select the statements relevant to this study. For the analysis of the selected statements, the study employs the framing approach introduced by Entman (1993), which is expanded by additional analytical categories that appear relevant for the complete capture of the discourse under study.

In the following chapter, the context of the political discourse in Ukraine is first introduced, explaining the discussion of the occupied territories in Ukrainian public discourse and also changes in media consumption in Ukraine. Chapter 3 explains the selection of actors and the compilation of the sample. Chapter 4 explains how framing is used as a research approach for this study, and chapter 5 outlines the detailed methodology of the study. Chapter 6 presents and evaluates the results of the analysis. The last chapter summarizes the characteristics of the examined discourse and the effects of its frames.

Context of the current political discourse in Ukraine

Occupied territories and their residents in Ukrainian public discourse

Despite the fact that the issue of "war" has dominated Ukrainian media coverage since the large-scale Russian attack, temporarily occupied territories of Ukraine are addressed in only 3,25 per cent of Ukrainian media content (Holub 2023). Based on the results of observations by the Institute of Mass Information (2023), the topic of occupied Crimea clearly takes the highest share of media coverage among the occupied territories, accounting for 66 per cent. The same report also reveals that the media they observed rely most heavily (approximately 40 per cent) on official Ukrainian authorities (e.g., the General Staff, and the Ministry of Defence) for reporting on the occupied territories. Approximately 12 per cent of citations are from Russian and foreign sources, as well as information from social networks, and in 9 per cent of cases, experts are also quoted. This indicates that the Ukrainian media generally describes the situation in occupied territories "from the outside" for Ukrainian citizens.

The Institute for Mass Information (2023) also provides an overview of the topics addressed in connection with the occupied territories. According to this overview, the media reported most frequently on the attacks by the Ukrainian Armed Forces on the Russian Black Sea Fleet (30 per cent) and on objects in the occupied territories (23 per cent). In comparison, topics directly concerning the representation and the perception of the residents of the occupied territories are mentioned much less frequently. For instance, ten per cent of the coverage is devoted to persecution and oppression of the local population. In five per cent of all articles about the occupied territories, the topic of collaboration is addressed, and the smallest share (three per cent) is allocated to predictions of the recapture of the affected areas. This suggests that media reports on the situation in the occupied territories are focused on military actions so that important aspects of daily life and the long-term prospects of the affected population remain underrepresented.

Several surveys conducted in the non-occupied parts of Ukraine since Russia's invasion indicate that the abovementioned topic carries potential social tensions. According to results from Rating Group (2023: 22), among all studied population categories, residents of the "People's Republics of Donetsk and Luhansk" were perceived the most negatively by the surveyed Ukrainians. In contrast, their attitude towards the inhabitants of occupied Crimea was 28 percentage points more positive. Attitudes towards people who have remained in the occupied parts of Kherson and Zaporizhzhia since 2022 are extremely positive according to this survey (Rating Group 2023).

This implies that the narratives Putin used to justify Russia's attack as a "defence of people in Donbas" fostered a negative attitude toward those residents of the "People's Republics of Donetsk and Luhansk" among the rest of the Ukrainian population. Unlike individuals who became victims of Russian occupation only in 2022, residents of the "People's Republics of Donetsk and Luhansk" are more likely to be perceived as supporters and drivers of the current invasion. The fact that the occupied parts of the Donbas were forcibly alienated from Ukraine with the Kremlin's help in 2014 also affected the perception of a social divide between "us" and "them," resulting in a decrease in empathy and solidarity for both sides.

In addition to the perceptions of the population groups under study, considerations of future reintegration measures are targeted in this study. The "most sensitive" aspects of this issue relate to handling individuals who have received Russian passports and identifying the actions in occupied territories that should be considered "collaboration." For instance, survey results from NaUKMA (2023) indicate that despite a social distance from people in the occupied territories, most respondents believe that obtaining a Russian passport should not be treated as a criminal offense. On the other hand, professional activities in the public administrations and authorities set up by the occupiers are clearly viewed negatively by the respondents, and most support the unconditional dismissal of those involved.

This suggests that residents of the occupied territories are a more complex topic in Ukrainian public discourse than they are in the general Ukrainian narrative of the Russian invasion. This and the strong underrepresentation of the residents of the occupied territories in media coverage of the current war indicate a need to examine the discourse surrounding this population group more closely.

Changes in media consumption in Ukraine

Current media consumption in Ukraine is characterized by the fact that social networks have been gaining increasing importance as sources of information for Ukrainian citizens. As numerous surveys conducted since the onset of the Russian invasion indicate (Opora 2022; Opora 2023; USAID-Internews 2022; USAID-Internews 2023), the majority of Ukrainians prefer to stay informed about current news through social networks. This trend has remained largely unchanged for the past two years. For two consecutive years, Telegram has been at the top of this trend, serving as the primary source of daily information for approximately 70% of Ukrainian users, while Facebook (19%) and YouTube (16%) hold the second and third positions, respectively (Opora 2023; USAID-Internews 2023). The high popularity of Telegram channels as a news source can be explained by the fact that their users find them convenient and practical as they receive news in real time via their smartphones without having to search for news updates themselves (Ukrainian Media and Communication Institute 2023: 5).

As the current study focuses on textual statements from selected actors, only the content from official websites and posts on official Telegram and Facebook accounts were included in the sample.

Selection of cases and composition of the sample

Since the aim of this analysis is to understand the discourse that influences policy-making regarding the occupied territories and their residents, the current analysis covers a wide range of formal and informal actors who were part of the executive branch of the Ukrainian government under Zelensky for the period under study. The following criteria were considered when selecting the actors:

1. Formal or informal actors should hold an office or position for which they were appointed or hired by Zelensky and were in this position at the time of the Russian large-scale invasion.
2. The actor should participate in at least an advisory capacity in decisions relevant to the current research question.
3. The actor should be as well-known as possible in general political discourse or in the political discourse specifically related to the subject of this study (both the surveys conducted since 2022 and the number of followers on the official social media accounts of the actors were taken into account to capture the level of their popularity).

Therefore, the current sample consists of the most well-known and relevant representatives for this study from the Cabinet of Ministers of Ukraine (Shmyhal and Vereshchuk), the Presidential Office (Yermak, Podolyak, and Arestovych (until 2023)), the Commander-in-Chief of the Armed Forces of Ukraine (Zaluzhnyi), the regional executive (Kim), and President Zelensky. In the sample, Arestovych is the only actor who has not been actively involved in Ukraine's policy-making since 2023, as he resigned as an advisor to the Presidential Office in January of that year and later fled Ukraine.

The attitude of the Ukrainian population towards the selected actors was measured several times after 2022 through opinion polls, allowing for changes to be tracked in the perception of these actors by Ukrainian citizens. For instance, Zelensky (approx. 90 per cent in May 2022) and Arestovych (approx. 65 per cent in May 2022) were rated as the most trustworthy by Ukrainian citizens in the first months after the Russian invasion (NDI 2022). In February 2024, Zaluzhnyi (approx. 94 per cent) topped the trust ratings, while trust in Arestovych (eleven per cent) experienced the greatest transformation among all selected actors, dropping from the top to the lowest positions (KIIS 2024).

The sample consists of written statements (including texts, interviews, short comments, etc.) by the chosen actors, which were published on their official accounts. In selecting the statements, an attempt was made to cover as broad a timeframe as possible, so the current sample includes published contributions from February 24, 2022, to December 31, 2023. Based on the relevance of social networks for Ukrainian citizens, it was decided to limit the sources of suitable statements to official websites (if available) and official accounts on Telegram and Facebook. The selection of content-relevant statements was made using the following search words in Ukrainian, Russian, and English: "occupied territories," "occupation," "People's Republic," "Donetsk," "Luhansk," "Donbas," "DNR," "LNR," "Crimea," "annexation," "annexed," "integration," "reintegration," "ORDLO," "Russian passports," "passportization," "collaboration," or "collaborator." After intensive reading of all selected contributions, the sample was further refined so that only texts mentioning territories occupied by Russia remained in the sample. The texts in which, for example, the search words "Donbas" or "collaboration" appeared without explicit reference to the occupied territories were removed from the sample. For a detailed documentation of data collection see Lysenko (2024).

As this study focuses on direct statements by the selected actors regarding the occupied territories and their inhabitants, only their self-authored statements were considered for analysis, excluding any forwarded posts. During the compilation of the sample, duplicate publications of statements were observed in most cases. This means that if content appeared once, for example, on an actor's official website, it most likely was simultaneously published on Telegram and Facebook. For this reason, it was decided to ignore the source of the contribution (official website, Telegram, or Facebook) when forming the text corpora. In the case of duplication, the most comprehensive version of a statement (e.g., the complete version on the official website instead of only a shortened version published on Telegram/Facebook) was included in the sample. For the sample, written statements from the following actors were selected:

	Position	Number of statements in the sample	Number of codes
Volodymyr Zelensky	President of Ukraine	96	1318
Andriy Yermak	Head of the Presidential Office of Ukraine	25	234
Mykhailo Podolyak	Advisor to the Head of the Presidential Office of Ukraine	15	166
Oleksiy Arestovych	Former Advisor to the Head of the Presidential Office of Ukraine	16	134
Valerii Zaluzhnyi	Former Commander-in-Chief of the Armed Forces of Ukraine	4	32
Denys Shmyhal	Prime Minister of Ukraine	7	60
Iryna Vereshchuk	Deputy Prime Minister of Ukraine and Minister for the Reintegration of Temporarily Occupied Territories	22	194
Vitaliy Kim	Head of the Mykolaiv Regional State Administration	8	74
Total:		193	2212

Table 1: The composition of the sample

Framing as a research approach

In this study, the analysis relies on the concept of framing, which initially defines a framework that retains important elements and leaves unimportant elements outside. Contemporary framing studies are based on two fundamental framing concepts from different disciplines: Bateson's definition of psychological frames (1972) and Goffman's (1974) micro-sociological definition of frames as a basic sociocognitive mechanism. Since the mid-1990s, frame analysis has also been strongly represented in the field of communication studies (López-Rabadán 2022: 1).

The advanced and expanded framing approach developed by Robert Entman (1993) proved to be most suitable for this study. Entman defined framing as a tool for the targeted representation of events that are formulated in accordance with a specific political direction and used to intentionally steer the opinions and action tendencies of individuals as well as collective viewpoints in certain directions. Framing can reinforce certain opinions, increase the pressure for specific actions, and use events to promote political and ideological currents, making them socially acceptable (Entman 1993: 52). Entman (1993) describes framing as a "fractured paradigm" and establishes the fundamental features of the field by defining the framing process as a strategic action that encompasses four discursive functions: "Problem Definition" (specific issues and problems addressed in the texts), "Causal Analysis" (causal connections considered between events and actors' behaviours), "Moral Judgement" (evaluation of the described content from a moral or ethical perspective), and "Proposed Solutions" (solutions to the problem previously defined in the text, which are proposed or at least addressed in the texts).

When applying the framing approach, however, it is crucial to consider the limitations of this concept. López-Rabadán (2022) emphasises that almost three decades after Entman's (1993) identification of a "fractured paradigm," framing studies still lack a unified theory on how frames are constructed and how they function precisely. The imprecise methodological criteria lead to a great variety of approaches, causing researchers to define and use frames in different or even contradictory ways (Matthes 2009). The lack of a unified theoretical foundation results in the collection of disconnected empirical data without integrating the findings into a common theoretical basis (D'Angelo 2012; López-Rabadán 2022). This affects the reliability and validity of the results and prevents the scientific standardisation of this research area (López-Rabadán 2022). Nevertheless, framing remains a central research approach in political communication studies as it provides deep insights into the mechanisms of information influence and public opinion formation (López-Rabadán 2022; Matthes 2014).

The essential epistemological assumptions of the frame approach include the intersubjective creation of meanings, the shaping of reality by discourses, and the centrality of context for meaning-making (Mendonça and Simões 2022: 345). In contrast to the concept of agenda-setting, which represents thematic priorities ("the what"), the concept of framing describes the strategic presentation of events ("the how") (López-Rabadán 2022: 4). De Vreese and Lecheler (2012) argue that framing as a theory connects three important areas: the strategic production of content, the characteristics of the message, and its individual, social, or cultural impact. This study focuses primarily on the examination of the characteristics of political actors' messages. To reconstruct frames from such messages, Van Gorp (2007) suggests paying attention to three aspects: framing devices, reasoning devices, and the implicit cultural phenomenon. Based on this suggestion, this study analyses which words, metaphors, and examples the actors use (framing devices), what kind of explanations or justifications they use in their argumentation (reasoning devices), and what their message reveals about the social and cultural values of the examined discourse (implicit cultural phenomenon).

Mendonça and Simões (2022: 346) also identify three ways in which frame analysis can be operationalised: 1) as sequence analysis of interactions, 2) as mapping of public discourses, and 3) as examination of framing effects. For the current analysis, the approach of mapping public discourses seems most suitable, as frame analysis attempts to capture and describe the characteristics of the official discourse about the residents of the occupied Ukrainian territories. The following analysis will also consider four basic principles of the framing approach by Entman (1993): ambivalence, selection, consistency (frames consist of coherent elements: problem definition, causal diagnosis, proposed solutions, moral judgement), and competition (Matthes 2014: 20-21). For this study, this means that the analysis will focus on how different political actors, competing for public dominance, strategically highlight or downplay specific aspects of issues. It will also aim to reconstruct coherent frame elements in the statements of these selected actors.

Methodology of data analysis

For the analysis of the texts and their contents, a computer-assisted discursive approach using the qualitative data and text analysis software MAXQDA (VERBI Software 2024) was selected. In the first step, groups of texts were created according to their authors, and in the next step, they were coded. An inductive approach was used for coding as this method allows for the identification of specific and previously undiscovered patterns in the discourses, thereby gaining a deeper understanding of the examined contexts (Matthes and Kohring 2008). In the context of a framing analysis, according to Entman (1993; 2007), selected text statements are coded according to Entman's four framing categories: "Problem Definition", "Causal Analysis", "Moral Judgement", and "Proposed Solutions". It is also important to note that the statements may not necessarily include all four of the categories described above. The categorical composition of the statements should be analysed to identify the possible intentions of their authors in the actual political context.

Applying an inductive approach for coding, the coding system, based on the framing categories according to Entman, was further expanded to include additional code groups relevant for this study: "Relevance of Occupied Territories in the Statement" (whether the occupied territories are mentioned as a side aspect in passing or explicitly addressed), "Recipients of the Message" (to whom the respective actors address their statements), "Reference to a Specific Territory" (whether the authors comment on the occupied territories in general or refer to specific territories), "Language of the Message" (which language or languages were used to write the statement) and "Emotional Message" (which emotions surround certain statements by the actors). Importantly, "Emotional Message" does not capture emotions in the sense of subjective states of feeling that occur in response to certain events or situations but rather emotional messages that are conveyed through the statements of the actors. The crucial difference from emotions is that an emotional message does not necessarily express the speaker's explicit emotional state but rather conveys a feeling directed at a recipient, with the aim of provoking an emotional response.

After the texts are coded, the most frequent subcodes are identified and analysed in relation to each other using mixed methods in MAXQDA. In the final step, the identified frames are described and analysed within the general political context of Ukraine.

Data analysis

What territories are discussed?

While analysing the statements, the first step was to examine which occupied territories were mentioned by the selected actors and which terms were used for them. In the entire sample, Crimea was mentioned most frequently (see Graphic 1). Crimea clearly dominates the statements of most analysed actors (Zelensky, Podolyak, Vereshchuk, Yermak, and Arestovych) and appears in at least a quarter of the statements of the remaining actors (Zaluzhnyi, Kim, and Shmyhal). The emphasis on Crimea could be related primarily to the fact that the return of Crimea is deeply associated with victory over Russia in Ukrainian political discourse. The motto "It started with the loss of Crimea, and it ends with its reclamation" has become symbolic as a necessary goal to end the war.

After Crimea, the actors most frequently refer to all occupied territories in general, without specifying exact regions. The frequency of this reference can be explained by the fact that it is well suited to articles with a general war context, in which authors do not need to be specific to make readers feel addressed. Notably, the territories occupied after 2022 are mentioned even less frequently than the Donbas territories occupied since 2014. These two categories are characterized by the greatest variance in the sample, and each has over six subcategories with specific territorial names; the lesser extent of the newly occupied territories in publications thus cannot necessarily be explained by a lower social relevance. Instead, these areas are most likely covered by the general category "All occupied territories." This indicates that authors prefer to speak in general terms rather than specifically about the topic of occupied territories.

In the examined discourse, another abstract category was also identified, which the actors used least frequently: the occupied Donbas and the South as identified without a clear definition of the temporal and territorial framework. It can be assumed that the South also includes Crimea and that the Donbas refers to areas occupied since 2014 and after 2022. In all (but one) of the cases, Crimea and all occupied territories are mentioned most frequently by the selected actors. The exception is Kim, who most often refers to the territories occupied after 2022. This can be explained by the fact that Kim is a regional politician (Head of the Mykolaiv Regional State Administration) and therefore speaks most about his area of responsibility, as his references to the Mykolaiv oblast stand out significantly.

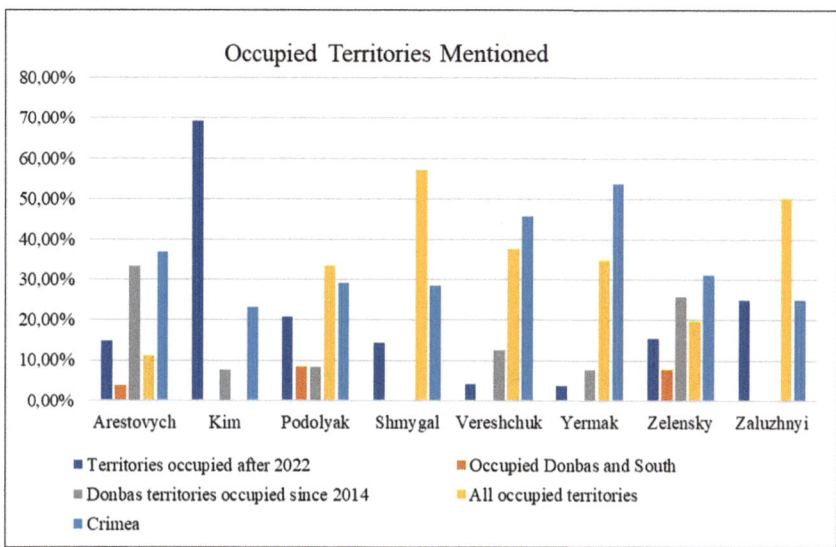

Graphic 1: *The mentioning of the occupied territories in the statements of the selected actors*

Zelensky offers the most balanced thematic coverage of all territories in his publications, although the majority of his statements also concern Crimea. The Donbas areas occupied since 2014 are mentioned by Arestovych most frequently, while in the statements of Shmyhal and Zaluzhnyi, these territories are not mentioned at all. In contrast, Iryna Vereshchuk, Minister for the Reintegration of Occupied Territories, has addressed the territories occupied since 2014 in only 12,7 per cent of her statements, even though these areas represent one of the greatest challenges for her ministry. The majority of Vereshchuk's statements focus on Crimea, which is mentioned more frequently by only Yermak. The analysis also makes it clear that the topic of Crimea is discussed in the most detail in the selected statements, while the remaining territories are mostly mentioned in passing or in a list (see Graphic 2). The issue of Crimea is most often the subject of an entire or half of a statement, whereas the Donbas areas occupied since 2014 are most frequently mentioned in passing.

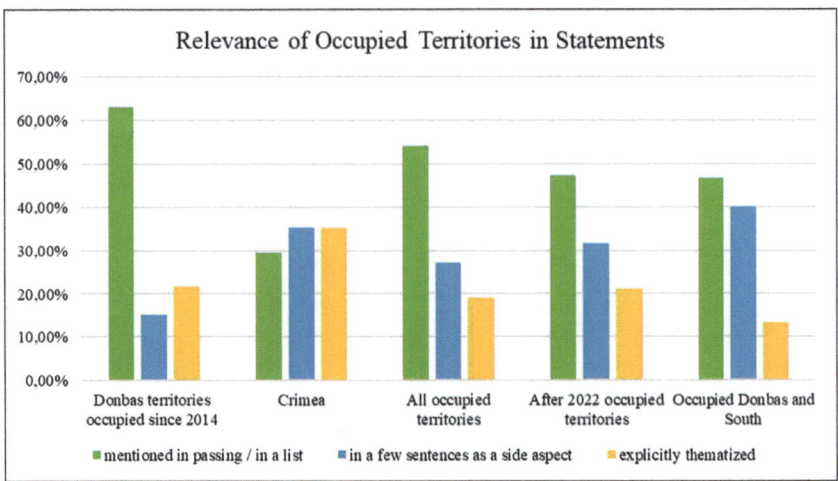

Graphic 2: Relevance of Occupied Territories in the statements of the selected actors

Who is addressed how and in which language?

Not only the occupied territories themselves are referred to in mostly general terms in the analysed statements, but the recipients of the messages are constructed in a similar way. In this sample, the actors appeal most often to all Ukrainians in general, without specifying their place of residence (see Graphic 3). The actors appeal next most frequently to other countries, which in this study refers to the governments and populations of all countries other than Russia and Ukraine, while the residents of the occupied territories themselves are most rarely addressed directly. Zelensky most frequently addresses all Ukrainian citizens, explicitly referring to them as Ukrainians (e.g., "Dear Ukrainians", "Great Ukrainian people"): *"Wishing you good health, dear Ukrainians!"* (Zelensky on his official website, 26.02.2023). Notably, the actors address the recipients in Russia even more frequently than the residents of the occupied territories.

Similar to the low mention of the occupied territories, the underrepresentation of appeals to recipients in occupied territories cannot be solely explained by a lack of interest in this group. It is important to note that the selected actors may not separate this population group from the category of "All Ukrainians"; therefore, "All" may also refer to the inhabitants of the occupied territories.

Based on the research question, this study focuses only on two kinds of directly related population groups: all Ukrainians and the residents of the occupied territories. Among the selected actors, Vereshchuk most frequently addresses the residents of the occupied territories (see Graphic 4). This can be explained by her position as Minister for Reintegration of the Occupied Territories. The proportion of statements from other actors addressing this population group ranges from

10 to 22,2 per cent. Notably, this group is entirely absent from the statements of Zaluzhnyi and Shmyhal.

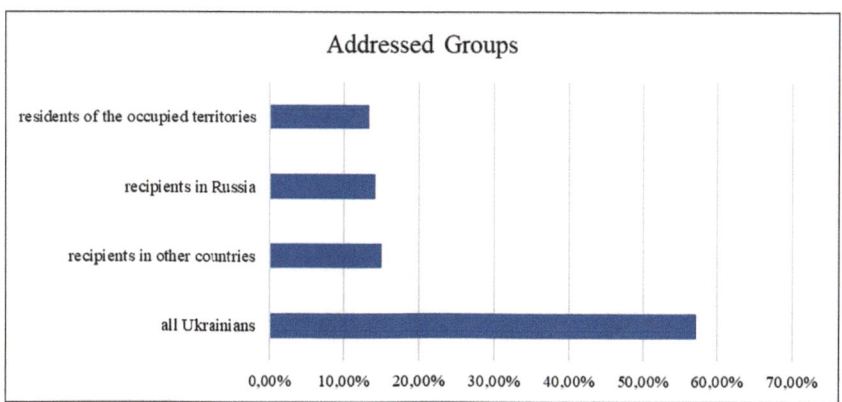

Graphic 3: Addressed population groups of the occupied territories in the statements

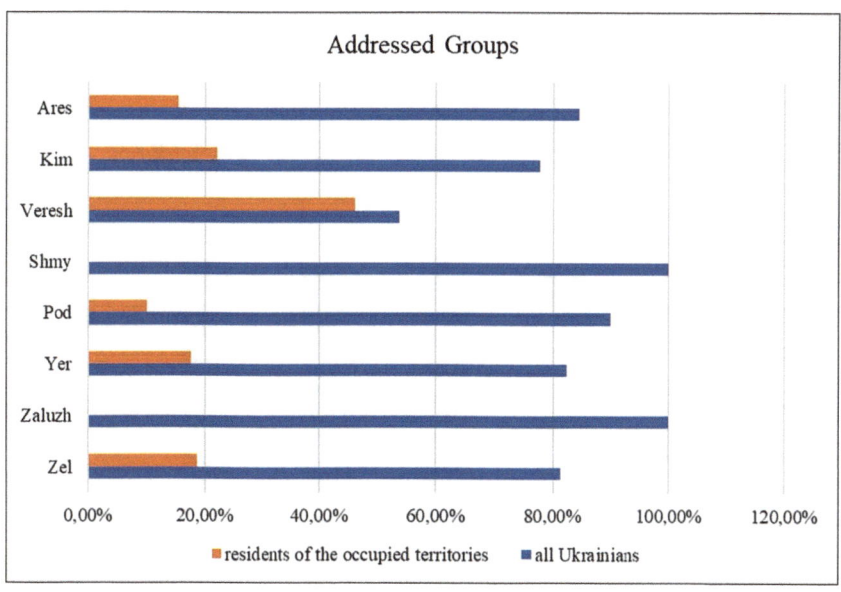

Graphic 4: Addressed population groups of occupied territories in the statements of each actor

One might assume that when appealing to Ukrainians in general, the selected actors mostly use the Ukrainian language, while when appealing to the residents of the occupied territories in particular, they tend to use the Russian language more often. This tendency, although not strongly pronounced, was observed only in the statements from Zelensky and Arestovych (see Graphics 5 and 6). The language use of Zaluzhnyi, Yermak, and Kim does not change with respect to the presumed recipients, while Podolyak and Vereshchuk use Russian when addressing the entire Ukrainian population in their statements. In appeals to the residents of the occupied territories, they exclusively use Ukrainian. This language choice contradicts the original assumption of linguistic preferences.

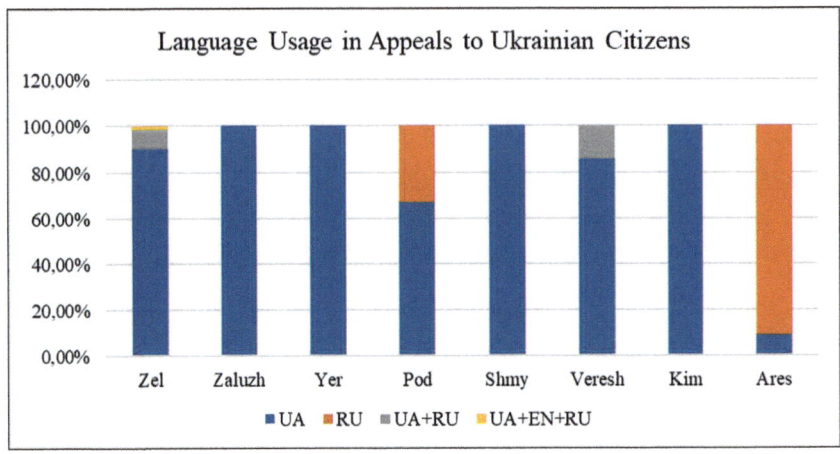

Graphic 5: Languages used in statements when appealing to Ukrainian citizens

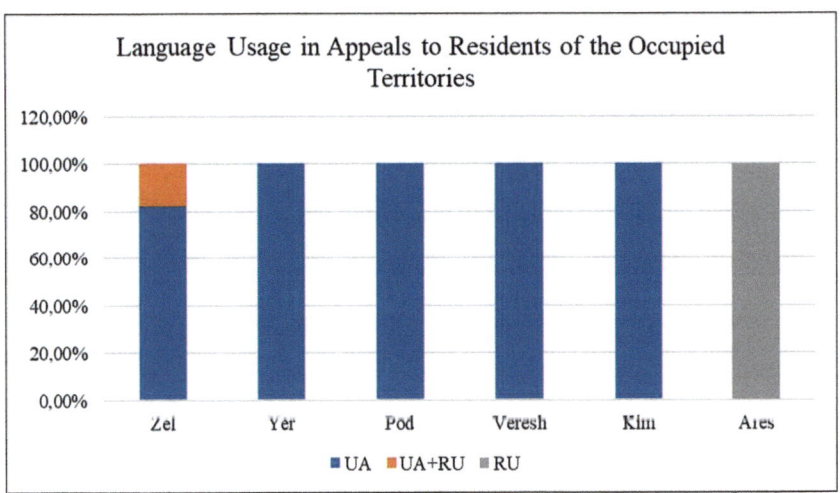

Graphic 6: Languages used in statements when appealing to residents of the occupied territories

In addition to language use, the emotional messages sent by the actors to the selected groups were studied (see Graphic 7). In appeals to all Ukrainians, two emotional messages stand out: confidence in victory and contempt. In contrast, appeals to the residents of the occupied territories are most often accompanied by admonitions and confidence in victory.

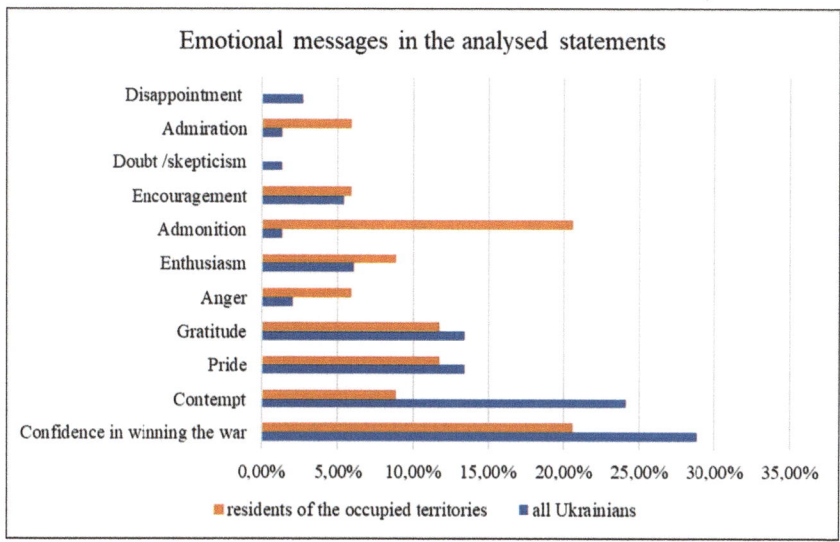

Graphic 7: Emotional messages expressed by the authors of the statements

It can be assumed that by expressing confidence in victory, the authors aim to encourage their recipients to continue believing in Ukraine's victory and to avoid falling into despair. This is because a depressed mood among the population may impact the course of the war, leading to loss of motivation, apathy, and personal disengagement. The actors in most cases try to strengthen the fighting spirit of the population and make it clear that their current suffering is not in vain and will eventually have a good end if they persevere. Shmyhal, Yermak, and Zelensky most frequently resort to this message, while Vereshchuk uses it the least among all actors. In the analysed statements, confidence in victory is primarily expressed through the use of modal adverbs such as "surely" and "clearly," which convey some certainty or clarity and are used in the texts to emphasize the speaker's conviction or assurance. Another feature identified in this emotional message is the use of future tense verbs, e.g., "we will win," "we will reclaim everything," without resorting to the subjunctive mood or modal verbs that would somewhat relativize the expressed intentions. Moreover, confidence as an emotional message is also expressed through sentences in the present tense with the modal adverbs "always" or "never," e.g., "Ukraine always comes back". This helps to reinforce the idea that the expressed content is a usual course of events that cannot be otherwise, e.g.: *"Crimea is Ukrainian territory. Always has been and always will be."* (Shmyhal on Telegram, 09.11.2023).

Contempt, as the second most frequent emotional message sent to all Ukrainians, is significantly less represented in statements to the residents of the occupied territories. Notably, contempt is directed not at the Ukrainian citizens but most often at the aggressor state (Russian government, soldiers, and citizens) and those who collaborate with it to create a feeling in the recipients of moral superiority over the aggressor. Contempt as an emotional message is particularly evident in the statements from Podolyak, Kim, and Arestovych and is not present in the statements from Zaluzhnyi and Shmyhal. A typical example of "contempt" in the analysed statements is the use of condescending remarks or insulting expressions that demean and portray the despised side as inferior, e.g., "z-creatures," "ru-devils," etc.: *"...Russia, which is a 'civilisational thug,' considers this an undeniable sign of weakness."* (Podolyak on Telegram, 09.06.2022). Furthermore, derogatory adjectives such as "ridiculous" or "pathetic" are often used to express contempt. However, in the sample, contempt is also expressed without derogatory statements, conveyed solely due to the context. For instance, the consequences of violent shelling can be described so that the main sentiment is clear even without insulting vocabulary.

To reinforce the feeling of contempt towards the aggressor, the actors often resort to irony and sarcasm, which can express a negative attitude without using demeaning expressions, for example, through the use of quotation marks. To express contempt, Arestovych and Podolyak often use these communication tools. They are also useful in expressing pointed criticism of certain actions indirectly, placing oneself above the despised individuals, e.g.: *"The 'second army of the world' is being wiped out in Donbas."* (Arestovych on Facebook, 30.06.2022). Those who handle irony and sarcasm well may be perceived by recipients as particularly intelligent and witty. Moreover, pointed and witty statements tend to remain longer in recipients' memories, as they stand out significantly among a mass of dry, unoriginal expressions. However, if this communication tool becomes too dominant, authors risk appearing arrogant, thus leading to more aversion in recipients than sympathy.

The use of admonitions as emotional messages, mainly directed at residents of the occupied territories, suggests that the Ukrainian government views this group as potentially more susceptible to persuasion to collaborate. The main function of admonitions seems to be to deter and dissuade people from cooperating with Russia. This emotional message is most frequently found in Vereshchuk's statements, while Zelensky, Zaluzhnyi, and Shmyhal do not use it at all. Admonitions were primarily identified through the use of imperative sentences and exclamations. These are used in statements both with negatives (e.g., "Do not go to the referendum") and without negation (e.g., "Leave the territories"). This emotional message is also strengthened by moral evaluations of potential actions, which the speaker warns the recipients about, e.g.: *"Firstly, it is illegal. Secondly, it is immoral."* (Vereshchuk on Telegram, 2.02.2023). This category often involves naming and discussing potential consequences that could occur if the recipients carry out the actions that the speaker admonishes them against.

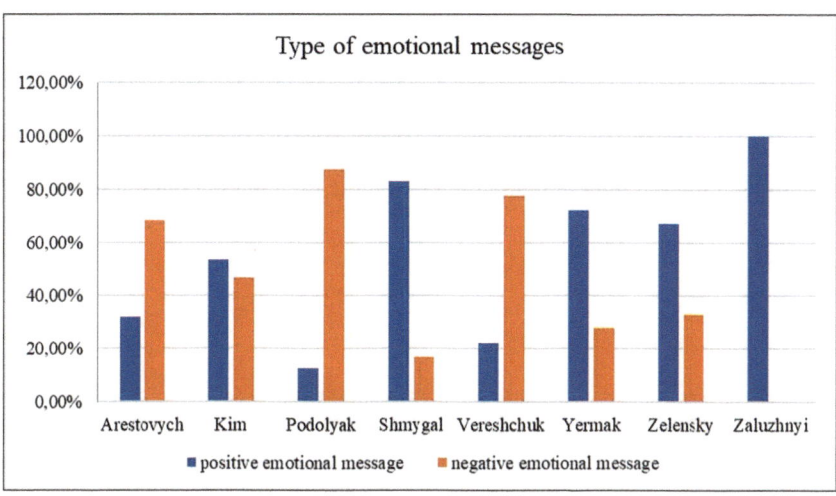

Graphic 8: Types of emotional messages expressed by the authors of the statements

For a better overview of the emotions that actors aim to evoke in their recipients, the emotional messages were categorized into two groups: negative (disappointment, doubt/scepticism, admonition, anger, and contempt) and positive (admiration, encouragement, enthusiasm, gratitude, pride, and confidence in winning the war) (see Graphic 8). According to this analysis, the leaders in expressing negative emotional messages are Vereshchuk, Podolyak, and Arestovych, while Zaluzhnyi, Shmyhal, Yermak and Zelensky focus on positive emotional messages. In this case, Kim gives statements representing the most balanced emotional palette among all actors.

Structure of the discovered frames

The analysis of framing categories in the statements showed that in the entire sample, the categories of "Moral judgements" and "Proposed solutions" are the most frequent and most strongly shape the discourse (see Graphic 9). The "Problem definition" category occurs less frequently, while the "Causal diagnosis" category is significantly underrepresented in the entire sample. Causes are most frequently discussed in Arestovych's statements, while other actors rarely address them. Notably, in most cases where the causes of the addressed problems were not named, moral judgements or proposed solutions compensated for the absence of this component. Due to the low relevance of the "Causal diagnosis" category in the sample and its strong variance within the code group, it was decided not to include it in the subsequent detailed framing analysis.

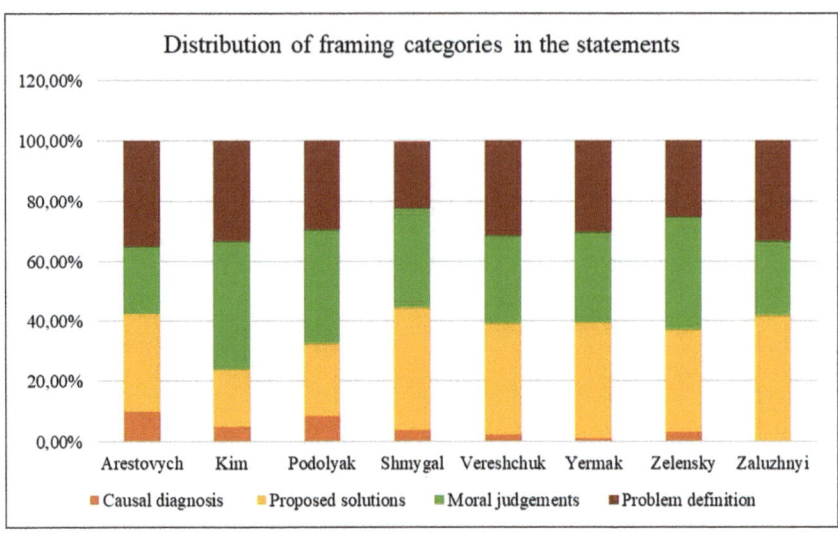

Graphic 9: Structure of the discovered frames

Moral judgement, as one of the most represented framing categories, is particularly influential in the statements by Kim, Podolyak, and Zelensky, while it occurs least frequently in the statements by Zaluzhnyi. Most moral judgements in statements from Zelensky and Podolyak concern the issue of Russian occupation and the situation in the occupied territories, while moral evaluations by Kim focus mainly on the problem of collaboration, e.g.: *"The enemy accomplice actively involved the city's residents in criminal activities."* (Kim on Telegram, 15.12.2023). Similar to the emotional messages, individual moral judgements were assigned to two groups (negative and positive), with the group of negative evaluations dominating the entire sample by a large margin (67,3 per cent). Positive moral evaluations predominate only in the statements by Shmyhal, while statements from the other actors contain few positive moral assessments.

With regard to negative moral judgements, accusations of cynicism are most frequently used, referring to Russia's military actions, its government and its army. Among the positive moral judgements, the emphasis on strength and fearlessness stands out, which in this sample is used exclusively in reference to Ukrainian citizens. The analysis also revealed that positive emotional messages most often accompany positive moral judgements, while negative emotional messages accompany negative moral judgements. Similar to the distribution of emotional messages in appeals to all Ukrainians, positive moral judgements are most often accompanied by the expression of confidence in victory, and negative moral judgements are accompanied by the expression of contempt (see Diagrams 1 and 2).

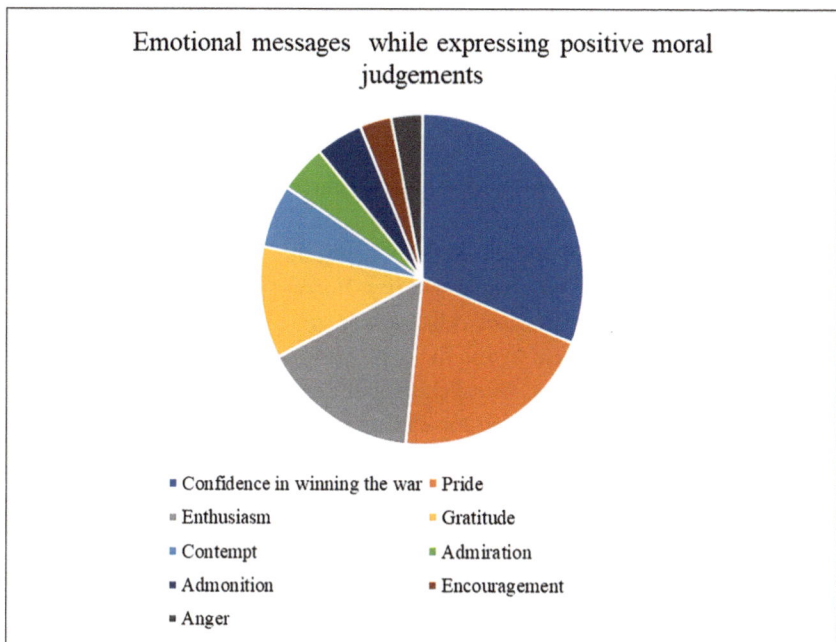

Diagram 1: Emotional messages expressed in positive moral judgements

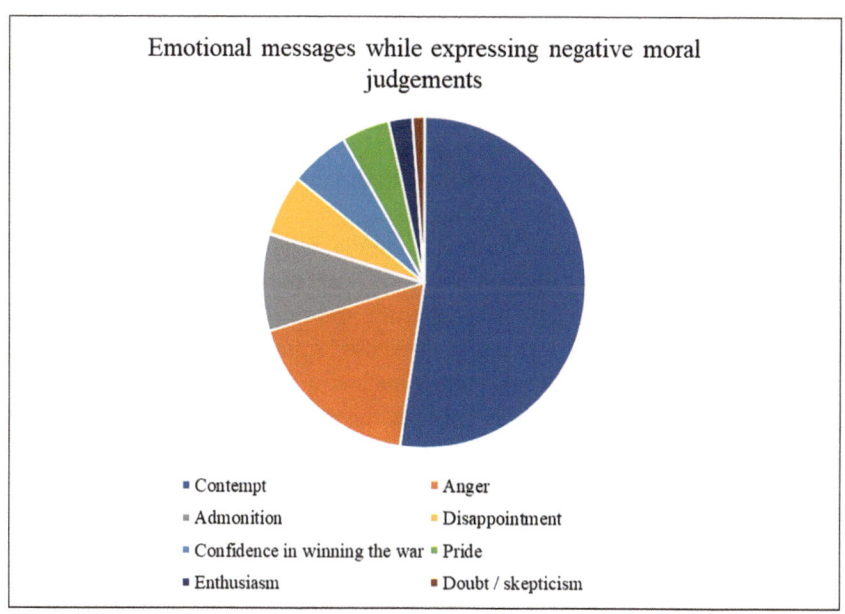

Diagram 2: Emotional messages expressed in negative moral judgements

While talking about the occupied territories, selected actors most frequently focus on the problems concerning the Russian occupation in general and the situation in the occupied territories (e.g., precarious humanitarian and economic situations, informational isolation, etc.), Russian military violence (e.g., Russia's strategy, shelling, destructions, etc.) and the activities of the occupation authorities (e.g., elections, issuance of Russian passports, etc.). One of the main characteristics of this discourse is that actors mostly limit their comments to defining the occupation by Russia as the main problem without going into detail about the situation of the people living in the occupied territories. In addition to the aforementioned problem definitions, the issues of collaboration (most frequently discussed by Kim, Podolyak, and Vereshchuk) and the oppression of people (almost exclusively in the statements by Shmyhal, Vereshchuk, and Zelensky) were also identified as frequent subjects. Other identified topics that appeared most rarely (less than 5 per cent of all codes related to this category), were not considered in the analysis.

Despite a general tendency in the entire sample, the distribution of addressed problems varies by the actor. The topics mentioned by Zelensky, Vereshchuk, and Yermak align most closely with the overall pattern observed in the sample.

Regarding the three most frequently discussed problems, the most common proposed solution is to continue fighting and to liberate the occupied territories using military means. In most cases, appeals are generally made to all Ukrainians. In the case of collaboration, actors most frequently suggested punishment for collaborators and advised abstention from actions that could be interpreted as collaboration under Ukrainian law (Articles 111 and 112 of the Criminal Code of Ukraine). Concerning the problem of the oppression of people, in addition to the aforementioned suggestions, the protection of the Crimean Tatars, the indigenous people of Crimea, is notably proposed. However, despite the wide range of proposed solutions, the idea of liberating Ukrainian territories by military means dominates among all studied actors.

Since future reintegration measures for the population from the occupied territories are of particular interest for the current study, the proposed measures that could specifically target this population group were analysed depending on the territory addressed. Among all identified solutions in the sample, three suggestions were determined to be the most relevant for the future of the occupied territories and their residents: reintegration, the reconstruction and upgrading of the occupied territories, and punitive measures for collaborators.

However, the analysis of these proposed solutions did not reveal a clear, unified action strategy for future measures concerning the occupied territories. Reintegration, the most frequently mentioned suggestion, is primarily proposed in relation to Crimea and less often in the context of all occupied territories (see Graphic 10). While discussing the other territories, reintegration measures are not mentioned at all. Measures for the reconstruction of the occupied territories are suggested for all examined areas except those occupied after 2022. Punitive measures for collaborators, however, are not addressed when discussing the Donbas regions occupied in 2014 and the occupied Donbas and South.

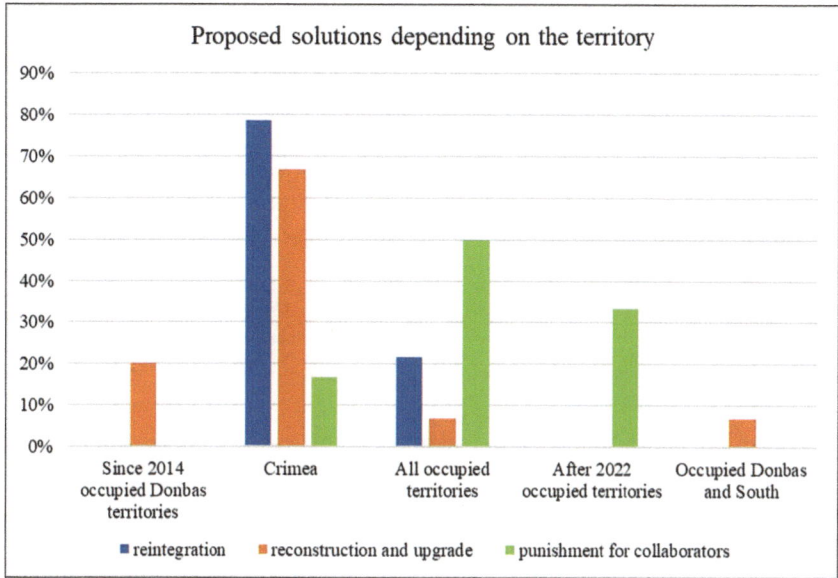

Graphic 10: Proposed solutions by the actors depending on the type of occupied territories

The fact that reintegration and reconstruction measures for the occupied territories are only mentioned by Zelensky, Vereshchuk, and Shmyhal, while punishment for collaborators is only proposed by Podolyak, Vereshchuk, and Arestovych, indicates that the studied discourse is characterized by weak coherence among proposed solutions. This creates the impression that there is no unified approach to communication or any action strategy regarding future reintegration among the selected actors.

Conclusion: Characteristics and effects of the analysed discourse

The current study explored how and to what extent official Ukrainian discourse addresses the issue of the residents of the occupied territories. The analysis aimed primarily to identify a clear communication and action strategy from the Ukrainian executive branch regarding this population group. Therefore, the discourse concerning the residents of the occupied territories was examined through the analysis of statements from formal and informal actors who, as part of the Ukrainian executive branch under Zelensky, were prominent in Ukrainian public discourse during the studied timeframe. Due to the increased relevance of social media as a news source for Ukrainian citizens, the sample consists of written statements from eight selected actors (Zelensky, Yermak, Podolyak, Arestovych, Shmyhal, Vereshchuk, Kim, and Zaluzhnyi) published on their official Telegram and Facebook accounts and on their official websites. The sample included 193 written statements, which were analysed using a computer-assisted discursive approach via MAXQDA.

The analysis was based on the framing approach suggested by Entman, examining the selected statements for problem definitions, causal diagnosis, moral judgements, and proposed solutions. To fully capture the nuances of official discourse, Entman's framing approach was expanded by several aspects in this study: reference to a specific territory, the relevance of occupied territories in statement, recipients, language, and the emotional message of the statement.

The analysis identified the following main characteristics of the official discourse and its possible effects on the recipients:

1. Despite the high variance in the addressed topics and of the communication tools used in their statements, the selected actors aimed to convey confidence that the occupied territories and their residents would once again become part of Ukraine. Notably, the actors clearly distance themselves from lamenting or speaking as a victim and instead try to speak from a position of moral strength. Consequently, moral evaluations clearly dominate the sample, primarily serving to depict Russia and its allies as morally reprehensible, in contrast to a brave and fearless Ukraine. The most frequently proposed solution, to continue fighting and to liberate the occupied territories by military means, also indicates that the focus of this discourse is on strength and encouragement.

2. The causes of the addressed problems are not elaborated upon in most cases, but the absence of this attribution is compensated for by moral judgements or solution proposals. Since the main problem in most cases is defined as the Russian occupation and Russia's aggression in general, the omission of any statement about causes creates the impression that, from the actors' perspective, the Russian occupation does not have serious or discussion-worthy causes and is self-explanatory through the moral reprehensibility of the aggressor.

3. The residents of the occupied territories are rarely addressed as a specific population group in this sample. It remains unclear whether the minimisation of messages to this population group is a deliberate strategy to avoid separating this population group from the rest of the Ukrainian population or merely a flaw in the official communication strategy, where they are not perceived as a target audience by the actors. Investigating the reasons for such a constellation of the official discourse is beyond the scope of this study, but it provides a framework for further research.

4. Official Ukrainian discourse contains significantly more discussion about the occupied territories and their reclamation than about their inhabitants themselves. This could provide fertile ground for provocative claims already being spread by Russian state media that Ukraine wants to reclaim its occupied territories even at the cost of numerous lives. In the sample, only Zelensky seems to have recognized this communication problem, as he counters this implicit accusation in his statements by emphasizing that "the fight continues for people, not territories, and that for Ukraine, unlike Russia, every human life is valuable".

5. Despite the observed tendency to speak in general terms about the Russian occupation, mentions of Crimea dominate the discourse on the occupied territories both in terms of frequency and in detail. Moreover, Crimea is the only occupied territory for which some specific reintegration measures after the end of the war were mentioned. Since the reclamation of Crimea is strongly associated with the end of the war, the specifically developed and communicated action strategies could reinforce the population's hope for success in this mission. However, this imbalance in the discussion of the territories may raise questions among readers about the relevance of the other occupied areas or if there is any plan for their reclamation and reintegration. This effect is further amplified by the examined solution proposals, which mainly relate to Crimea.

Notably, the main focus of the statements on the residents of the occupied Crimea is on the Crimean Tatars. The repressive measures problematized in the sample have the same focus, which concerns almost exclusively the oppression of Crimean Tatars. The same imbalance, as in the case of the ratio of mentions of the occupied territories, could cause the remaining resident groups to feel overlooked.

6. Concerning the language use in the statements of most actors, there does not seem to be a communication strategy aimed at addressing a specific recipient group in the language they presumably use in their everyday communication. However, it is evident that in some cases, Zelensky adapts his use of language to some targeted groups of recipients. For example, Zelensky uses Russian when appealing to the citizens of Russia or English when addressing foreign citizens or governments. For the residents of the occupied territories, who, according to survey results before the start of Russian aggression in 2014 and 2022, described themselves as predominantly Russian-speaking, this linguistic strategy of matching language to audience does not appear to be applied by all studied actors.

7. Zelensky's position, which most significantly shapes the official discourse due to the high amount of coded content, is characterized mainly by his general avoidance of socially controversial topics regarding the residents of the occupied territories. He mostly speaks on this topic in the broader context of Russian aggression, while sensitive issues are more frequently addressed by other actors (Kim, Podolyak, Vereshchuk, and Arestovych). This creates the impression that

Zelensky deliberately refrains from engaging with critical topics concerning the residents of the occupied territories to maintain a positive and unifying impact on the recipients. The desire to be perceived as positive by recipients can also be inferred from the fact that Zelensky is one of the few actors in the sample whose statements predominantly convey positive emotional messages.

Although some tendencies can be identified in the current official discourse, it remains unclear whether these result from a deliberate communication strategy regarding the occupied territories and their residents or a natural inclination to avoid a highly sensitive topic. This could provide a basis for further research, exploring both the issue of the residents of the occupied territories in the broader political discourse in Ukraine (e.g., including opposition actors) and the reasons for the significant underrepresentation of this topic.

References

Bateson, Gregory. A Theory of Play and Fantasy. Boston: MIT Press, 1972.

Brugman, Britta C. / Burgers, Christian. "Political framing across disciplines: Evidence from 21st-century experiments." In: Research and Politics, April-June 2018, pp. 1-7.

D'Angelo, Paul. "Studying framing in political communication with an integrative approach." In: American Behavioral Scientist 56 (2012), pp. 353-64.

De Vreese, Claes H. / Lecheler, Sophie. "News framing research: An overview and new developments." In: The SAGE Handbook of Political Communication (2012), pp. 292–306.

Entman, Robert M. "Towards Clarification of a Fractured Paradigm." In: Journal of Communication 43/4 (1993), pp. 51-58.

Entman, Robert M. "Framing Bias: Media in the Distribution of Power." In: Journal of Communication 57 (2007), pp. 163-173.

Goffman, Erving. Frame Analysis: An Essay on the Organization of Experience. Cambridge: Harvard University Press, 1974.

Holub, Olena (Institute of Mass Information): "Temporarily occupied territories: what Ukrainian online media write about them", 2023. Accessed 3 March 2024: https://imi.org.ua/monitorings/tymchasovo-okupovani-terytoriyi-shho-pro-nyh-pyshut-ukrayinski-onlajny-i57440

Kyiv International Institute of Sociology. "The course of affairs in the country and trust in political, military and civil society actors. Public opinion polls' results", 2024. Accessed 3 March 2024: https://kiis.com.ua/?lang=ukr&cat=reports&id=1368&page=1

López-Rabadán, Pablo. "Framing Studies Evolution in the Social Media Era. Digital Advancement and Reorientation of the Research Agenda." In: Social Sciences 11 / 9 (2022), pp. 1-19.

Lysenko, Yana. "Text corpus: Residents of the Territories Occupied by Russia in Official Ukrainian Discourse," v. 1.0, Discuss Data, 2024: https://doi.org/10.48320/BE472806-5F65-427B-8E2B-203779089F28

Matthes, Jörg / Kohring, Matthias. "The content analysis of media frames: Toward improving reliability and validity." In: Journal of communication 58 (2008), pp. 258-279.

Matthes, Jörg. "What's in a frame? A content analysis of media framing studies in the world's leading communication journals, 1990-2005." In: Journalism & Mass Communication Quarterly 86 (2009), pp. 349-367.

Matthes, Jörg. Framing. Baden-Baden: Nomos Verlagsgesellschaft, 2014.

Mendonça, Ricardo F. / Simões, Paula G. "Frame Analysis." In: Research Methods in Deliberative Democracy (2022), pp. 345-355.

National Democratic Institute. "Opportunities and obstacles on the way to Ukraine's democratic transition", 2022. Accessed 3 March 2024: https://www.ndi.org/sites/default/files/NDI%20Survey%20UKR.pdf

Opora. Report on the survey "Democracy, rights and freedoms of citizens and media consumption in the conditions of war", 2022. 3 March 2024: https://www.oporaua.org/viyna/sotsiologichne-doslidzhennia-demokratiia-prava-i-svobodi-gromadian-ta-mediaspozhivannia-v-umovakh-viini-24261

Opora. Report on the survey "Media consumption of Ukrainians: the second year of full-scale war", 2023. Accessed 3 March 2024: https://www.oporaua.org/polit_ad/mediaspozhivannia-ukrayintsiv-drugii-rik-povnomasshtabnoyi-viini-24796

Rating Group. "Sociological research for Independence Day: Perception of patriotism and the future of Ukraine", 2023. Accessed 3 March 2024: http://ratinggroup.ua/research/ukraine/soc_olog_chne_dosl_dzhennya_do_dnya_nezalezhno_uyavlennya_pro_patr_otizm_ta_maybutn_ukra_ni_16-20_se.html

Razumkov Centre. "Citizens' assessment of the situation in the country. Trust in social institutes, politicians, officials and public figures", 2023. Accessed 3 March 2024: https://razumkov.org.ua/napriamky/sotsiologichni-doslidzhennia/otsinka-gromadianamy-sytuatsii-v-kraini-dovira-do-sotsialnykh-instytutiv-politykiv-posadovtsiv-ta-gromadskykh-diiachiv-stavlennia-do-provedennia--zagalnonatsionalnykh-vyboriv-v-ukraini-do-zavershennia-viiny-veresen-2023r

School For Policy Analysis (NaUKMA). "Reintegration of de-occupied communities and social cohesion", 2023. Accessed 3 March 2024: https://spa.ukma.edu.ua/wp-content/uploads/2023/11/RG_ShPA_Reintegration_2000_102023-1.pdf

Ukrainian Media and Communication Institute. "How Non-Institutionalized News Telegram-Channels Operate and Capture the Audience in Ukrainian Segment. Analytical report." Kyiv: Ukrainian Media and Communication Institute NGO, (2023), pp. 1-69. Accessed 3 March 2024: https://www.jta.com.ua/wp-content/uploads/2023/02/Telegram-Channels-2023.pdf

USAID-Internews. "Ukrainian media, attitudes and trust in 2023", 2022. Accessed 3 March 2024: https://internews.in.ua/wp-content/uploads/2022/11/Ukrainski-media-stavlennia-ta-dovira-2022.pdf

USAID-Internews "Ukrainian media, attitudes and trust in 2023", 2023. Accessed 3 March 2024: https://internews.in.ua/wp-content/uploads/2023/10/Ukrainski-media-stavlennia-ta-dovira-2023r.pdf

Van Gorp, Baldwin. "The Constructionist Approach to Framing: Bringing Culture Back In." In: Journal of Communication 57/1 (2007), pp. 60-78.

VERBI Software. "Online Manual MAXQDA 2024", 2024. Accessed 3 March 2024: www.maxqda.com/help-max24/welcome

Living Through Narratives
A Psycholinguistic Study of War Stories
by Bohdan Lepky and Today's Ukrainians in Print
and Digital Media

Serhii Zasiekin

Abstract

This study is focused on identifying psycholinguistic markers of war-related trauma in the narratives of today's Ukrainians and of the well-known Ukrainian writer and public figure Bohdan Lepky. Our aim is twofold: (i) to identify any changes that occurred in the civilians at the end of the first year of Russia-Ukraine war; (ii) to compare the current data with those found in Lepky's literary works. Using the LIWC 2015 software (Pennebaker et al., 2015) and ANOVA statistics, we analysed 354 publicly available testimonies from the Facebook group "Writings from the War", together with 31 war stories by Lepky, and a reference corpus of 100 literary prose texts by Ukrainian authors. The social media narratives revealed three prominent LIWC categories related to social relations: 'We', 'Social' and 'Family'.

A greater prevalence of the categories 'Affiliation' and 'Achievement' in the Facebook corpus compared to the literary and Lepkyi corpora, implied that Ukrainians were praising their advances after one year of resistance and fighting against the enemy. Lepky's stories showed none of these features, implying that he could not rely on others during the tragic times of that war a century ago. The data also showed that the Facebook users opted for an analytical style, focusing more on cognitive reprocessing of their painful experiences, thus reinforcing psychological resilience, which is seen as a positive trait in terms of mental health (Pennebaker et al., 2000). The style of Lepky's war stories is more narrative as the Dynamic Index of his texts is significantly higher than in two other corpora.

Keywords

Psycholinguistic; war stories; narratives; prose text; Facebook

Not for the ruin was our birth, And not for the war or fight –
The country's good is of real worth, To bring our people light…
Bohdan Lepky

Introduction

The psycholinguistic study of this language can provide a researcher with a key to open the "black box" that is our mind, to learn what is going on inside, to explain individual behaviour and cognitions, and to predict their future actions. Stress and trauma, both individual and collective, are part of our emotional environment and accompany us throughout our lives, becoming particularly acute in the atrocities of war, in physical and moral pain, and in displacement. The war currently being waged by Russia against Ukraine is characterised by a number of indicators that point to the commission of genocide against the Ukrainian population. Research investigating various genocides including Holocaust and Holodomor along with the collective trauma "active forgetting" mnemotechnologies (Aydin 2017) has demonstrated the significance of transgenerational transmission of the collective trauma narrative (Zasiekina et al., 2021). Despite its deep culturally embedded nature, research conducted by American social psychologist James Pennebaker and his colleagues (1999) indicates that trauma can be reduced or alleviated when individuals verbalise their negative emotions and experiences in narrative form.

The years 2022-2024 have witnessed a growing trend of research into the cultural, linguistic, social and psychological consequences of the Russia-Ukraine war. No research has yet examined the psycholinguistic characteristics of war stories told by civilians during Russia's full-scale invasion of Ukraine. This psycholinguistic study is part of the broader project War in the Narratives of Bohdan Lepky and the Present-day Ukrainians, and is in line with our previous study (Zasiekin et al., 2022) of war testimonies posted on Facebook by Ukrainians during the first six months of the war. Our previous findings indicated that despite exposure to war atrocities and military trauma, authors tended to actively reappraise traumatic events, which could lead to the relief of traumatic stress (Pennebaker & Seagal, 1999).

The current investigation has two main objectives as it involves Bohdan Lepky (1872-1941), a writer of fiction. Firstly, to establish if the aforementioned tendency holds true for civilians one year after the war has started.

The style of narrating the experience in stories alongside psychologically and linguistically meaningful categories of individual war experiences by today's social media users is a marker of this tendency. Secondly, the current data will be compared with a corpus of war stories told by Bohdan Lepky (1872-1941) in print media. The Lepky's narratives depict similar content features of the tragedy of Ukrainians who had fought in opposite camps during World War I and later for their independence against the Russians a century ago, and today's war against the same enemy.

This study aims to explore how language reflects the psychological impact of war on humans and demonstrates potential methods for accessing individuals' thoughts through their written testimonies. The analysis of individual experiences of the war and its impact on mental health addresses the following research questions:

Research Question 1: What psycholinguistic categories are significant in individual experiences of the war reflected in Facebook posts and in Bohdan Lepky's stories?

Research Question 2: Which style of thinking, i.e. analytic or narrative, is predominant in social media and Lepky's narratives when representing war experiences?

Material

Facebook stories

The research material consisted of three corpora. The first corpus, 'Writings from the War', contains 354 publicly available written testimonies, comprising 78,952 word tokens, and was posted on Facebook platform (https://www.facebook.com/WritingsFromTheWar). The selection of this platform as a source of war testimonies was based on several factors. First, the accessibility of the stories was a significant consideration. Second, the cognitive, expressive, and emotional load of the stories, together with their length – the advantageous feature of Facebook compared to other social media – was taken into account. Third, the fairness in reproducing the events was a key factor. The 'Writings from the War' voluntary project was founded and is headed by Alex Averbuch from Edmonton, Alberta, Canada. Its aim is to gather, conserve, and circulate first-hand accounts of Ukrainians during the Russia-Ukraine war. Documenting numerous testimonials in the form of texts, videos, and photographs provides an unbiased and comprehensive insight into the tragedy of the war in Ukraine and allows the voices of those affected to be heard.

The testimonies come from a diverse group of individuals, including adults and children, internally and externally displaced citizens, rescuers, and the rescued. This allows anyone who wishes to share their experience during this catastrophe to be included and heard. This provides a comprehensive and diverse account of everyday experiences during wartime, reflecting private and personal perspectives.

All narratives included in this study were posted on Facebook from the beginning of the war in February 2022 to March 2023.

This study utilized two additional corpora described below.

Why social media?

The media plays a vital role today in shaping public opinion during wars and conflicts, both domestically and globally. The war in Ukraine serves as a clear illustration of how social media has become a significant battleground in the information war between Russia and Ukraine (Krylova-Grek 2022: 83).

The scale of Russian aggression and the speed of online updates, generated by millions of users, have compelled mainstream media outlets to rush and break stories that compete with the current pace of social media cycles, even before their development is complete. The value of legitimate open-source investigations aimed at debunking Russian claims has been undermined, resulting in the propagation of misinformation that benefits the Kremlin (Karalis 2024).

Russian capabilities in the information war have thrived due to the diversion of attention to other armed conflicts, e.g. between Israel and Gaza, Armenia and Azerbaijan, causing the emergence of so- called 'Ukraine Fatigue.' Russian disinformation can effectively permeate social media spaces under these conditions, significantly influencing global public opinion about the war in Ukraine. Russian disinformation can effectively permeate social media spaces under these conditions, significantly influencing global public opinion about the war in Ukraine. The tactics employed by Russian disinformation are deceptive. They use a brand of malinformation that shifts attention from the atrocities, murders of civilians, their tortures – that have features of genocide – to Ukraine's and Western faults, despite Western governments repeatedly expressing their commitment to help Ukraine withstand Russian aggression.

Therefore, it is crucial to be vigilant and cautious in the consumption of information and to support legitimate open-source investigations to debunk Russian claims (Karalis 2024). The stories of first-hand accounts shared by Ukrainians on social media help demystify the false information and propaganda disseminated by Russia, and are part of the national and global cultural memory of the Russian invasion of Ukraine.

Why Bohdan Lepky?

For almost a century after World War I, it was widely assumed that only soldiers who had experienced war at first hand could write novels about it. This belief was so deeply ingrained in the conventions of the genre that the authority of veteran authors was rarely questioned by writers, critics, reviewers or even scholars.

According to this perspective, war stories could only be recounted by veterans (Eisler 2022: 1). In 2016, the veteran author Matt Gallagher (2016) published an article titled *"You Don't Have to Be a Veteran to Write about War,"* challenging the unspoken rule. Gallagher pointed out the baseless idea that one shouldn't write about war unless they participated in it as a combatant or otherwise survived its destruction. He called this a silly notion and highlighted the power of fiction to create worlds from the imagination, regardless of the author's experience. It is important to note that writing is writing, and a writer who is skilled at the craft doesn't need to have experienced combat first-hand to take readers to war. As Eisler (2022: 1) argues, a skilled writer can use their imagination to create a realistic portrayal of war.

Bohdan Lepky, a non-combatant during the First World War, witnessed the sufferings of both military and civilian populations. He trustfully and profoundly exposed these experiences in his stories, making him the most prolific Ukrainian writer after Ivan Franko in terms of the scope of his work. At the outbreak of the First World War, he embraced his role as a historian and patriot. He recognized that the burden of the difficult and tragic times fell upon the intelligentsia, and that they must rise to the occasion and serve as an asset to the people in order to secure a brighter future for our nation. During the First World War, Ukrainians fought on opposite sides in both the Russian and Austro-Hungarian armies.

However, the sense of Ukrainianness gradually grew among them, thanks in large part to educators and spiritual leaders like Lepky. Back in November 1915, he was sent to work in German camps for Ukrainian prisoners of war in Rastatt and later in Wetzlar under the leadership of the Union for the Liberation of Ukraine (ULU). It is interesting to note that the ULU was founded in 1914 in Lviv and moved to Vienna in August of the same year, where Lepky joined the Union (Lanovyk & Lanovyk 2022). His poetic and prose writings in *The Bulletin of the Union for the Liberation of Ukraine* not only raised the national consciousness of the Ukrainian public, but also garnered the unanimous support of the entire Ukrainian soldiering.

Lepky's books draw many parallels with the worldview of Ukrainians in 2022-2024. The historical similarities between the present and the events depicted in Lepky's epic pentalogy *Mazepa* are striking. The book strongly supports the Ukrainians' resistance and opposition to the Russians from the Great Northern War (1700-1721), depicting the Moscovians' violence and atrocities, as well as the Red Army terror (1917-1921) and their 'extraneousness' (Isaiuk, 2022). His works, being "The Gate to Eternity" (Suproniuk, 2022), effectively convey the collective trauma of the Ukrainian nation as a part of Europe since its foundation during the time of Kyivan Rus'.

Baturyn's death is alluded to in Poltava. The descriptions of violence against the Ukrainian population in Baturyn in 1708 are comparable to the atrocities committed by Russian soldiers in Bucha, Mariupol, Vorzel, Irpin', Hostomel and other cities of Ukraine in 2022. This comparison highlights the historical precedent for current Russian actions and intimidation tactics.

It is worth noting that this is not the first time that history has repeated itself in new circles of historical reality. This historical parallel serves as a reminder that the current situation is not unique and that there are lessons to be learned from the past.

As Lepky portrays in his war stories the human psyche as 'a stream of consciousness' through first- person narration, this creative approach aligns his stories with the genre of *'I-person'* social media.

The Lepky's corpus contained his 31 war stories (Lepky 2011) published in print media, with a total of 41,028 word tokens.

Literary corpus

The third, reference, corpus compiled by Zasiekin (2020) and displayed on Mendeley (Zasiekin 2022), included 100 literary prose texts by classical and modern Ukrainian writers (19th-21st centuries) of 20,000 words each, totalling 2,000,000 tokens. Due to their balanced inclusion of narration, description and dialogue segments, they represent a comparable corpus of texts to the Facebook and Lepky texts.

The infrequent use of the war theme by the authors in the literary corpus serves as a 'norm', a point of reference for the other two corpora.

Tools

Patterns of language use are very valuable for the analysis of interactions because a significant amount of communication between individuals is conveyed through language. Linguistically, all texts are made up of words. All words in texts are either function words – prepositions, pronouns, conjunctions, interjections – or notional words – nouns, verbs, adjectives, adverbs. In terms of psycholinguistics, function words used in speech primarily reflect unconsciously *how* we communicate, i.e. our style, whereas notional words convey *what* we are saying. As a result, style words can be used as linguistic and psychometric indicators of something that is not immediately detectable from the speech -- they can measure the social and psychological experiences of the speaker (Tausczik & Pennebaker 2010). Therefore, this study employs a psycholinguistic analysis using the Linguistic Inquiry and Word Count – LIWC 2015 computerised tool (Pennebaker et al., 2015; 20-23) and a statistical method using ANOVA analysis with a significance level set at 5%. The statistical analyses, conducted in SPSS version 26.0, compared the *Writings from the War* and *Lepky's corpora* with the reference *Literary corpus* in terms of percentages of words belonging to specific linguistic and psychological categories, as well as indicators of narrative vs. analytical style.

LIWC calculates the percentage of words in the input text that belong to over eighty categories. The 2015 English version of the software contains over 6,400 words, grouped into categories such as grammatical, psychological, social relations, and personal themes (e.g. friend, female, male, work, religion, family, home, leisure, death, money). The programme is designed in such a way that when the engine detects a particular word in the text, it assigns it to one or more of the appropriate categories.

LIWC consists of two blocks: a processing block and dictionaries. Each word in the text is associated with a word from the matching dictionary, with function words and full-meaning words being distinguished for categorisation purposes. Function words, such as pronouns, articles, prepositions, and conjunctions, are distinct from semantic words, which include verbs, nouns, adjectives, and adverbs. It is important to note that while the average native speaker has a vocabulary of, say, about 100,000 words, only 0.5% of these are function words. Despite this, function words make up a significant portion of our everyday speech, comprising around a third (in Ukrainian) to a half (in English) of the words we use.

By using the established weights of these categories in the texts, similarities and differences between Lepky's century-old war stories and present-day war writings by civilians in Ukraine were identified.

The Categorical Dynamic Index (CDI) (Jordan & Pennebaker, 2017) defines the style of representation of the war experience as either analytical or narrative. This unique style combines abstract thinking with cognitive complexity. A lower CDI score indicates a higher frequency of auxiliary verbs, adverbs, conjunctions, impersonal pronouns, negations, and personal pronouns. Research on English texts indicates that certain categories of words, such as pronouns and auxiliary verbs, are commonly used in time-based stories and reflect a dynamic or narrative language style (Biber, 1988). For this study, which focuses on Ukrainian texts, we employed six function categories: prepositions, personal pronouns, indefinite pronouns, conjunctions, adverbs, and negations. The study employs the Ukrainian version of the LIWC 2015, which is the result of my close collaboration with its author and designers (see: Pennebaker et al, 2023).

As Ukrainian does not have articles, adjectives are used instead to modify nouns and also serve as markers of analytic thinking in the formula below (Pennebaker et al., 2014).

$$Categorical = (adj + prep)/2 \qquad (1)$$

We also modified the original English-text-based formula for dynamic index by deleting auxiliary verbs, since the Ukrainian language does not have such verb forms:

$$Dynamic = (ppron + ipron + conj + adverb + negate)/5 \qquad (2)$$

Results

Variables	Lepkyi's war narratives (n=31)		Facebook war narratives (n=354)		Literary texts (n=100)		F	p
	Mean (SD)	Min-Max	Mean (SD)	Min-Max	Mean (SD)	Min-Max		
Anger	.17 (.20)	.00 .76	.09 (.33)	.00 3.85	.19 (.11)	.04 .83	4.534	.011
We	.39 (.40)	.00 1.74	.72 (.95)	.00 6.12	.22 (.21)	.00 1.40	15.746	<.001
Social	2.90 (.79)	1.45 4.85	3.91 (2.99)	.00 33.33	3.12 (.56)	1.73 4.80	5.155	.006
Family	.43 (.27)	.00 1.04	.98 (2.20)	.00 33.33	.42 (.18)	.14 1.06	4.224	.015
Cogproc	8.63 (2.01)	5.37 13.09	6.48 (3.97)	.00 25.00	7.87 (1.35)	5.15 13.43	10.223	.001
Cause	2.31 (.71)	1.16 4.32	1.34 (1.48)	.00 16.67	1.61 (.46)	.79 3.94	8.883	.001
Affiliation	.59 (.45)	.00 1.74	1.12 (2.06)	.00 33.33	.52 (.24)	.21 1.71	5.316	.005
Achiev	.24 (.19)	.00 .71	.57 (.91)	.00 6.67	.34 (.13)	.09 .74	5.216	.006
Focusprese nt	1.32 (.75)	.00 3.33	1.42 (1.65)	.00 18.18	.98 (.33)	.39 2.12	3.787	.023
Sixltr	25.22 (4.77)	18.38 33.78	31.13 (10.50)	10.26 100.00	29.37 (3.95)	21.00 40.41	6.589	.002
Filler	.03 (.06)	.00 .26	.05 (.21)	.00 2.21	.14 (.12)	.00 .78	8.143	.001
Informal	.97 (.50)	.24 2.43	.76 (1.24)	.00 16.67	1.09 (.39)	.00 2.39	4.100	.017
Swear	.05 (.09)	.00 .31	.02 (.09)	.00 .89	.07 (.07)	.00 .45	13.670	.001

Table 1: Mean Percents, Standard Deviations and ANOVA Outcomes for Linguistic and Psychological Meaningful Categories of War Narratives on Facebook (n = 354), War Narratives by Lepkyi (n = 31) and Other Literary Texts (n=100)

Analysis of the three LIWC categories related to social relationships shows that they were more prominent in the Facebook corpus than in the other two literary corpora (see Table 1). Specifically, the significant role of "Affiliation" and "Achievement" in Facebook narratives is consistent with the greater impact of the "We" and "Social" categories, which enhance the effect of 'togetherness'. The categories of "Affiliation" and "Achievement" included in the LIWC 2015 program belong to the LIWC super-category of "Drives," which, along with "Power," pertain to three fundamental human needs (McClelland, 1961). In other words, the LIWC reflects the degree of vitality of these values for the author of the text under assessment.

Lepky's stories do not exhibit this trend, suggesting that he was unable to rely on others during the war's trying times. It is worth noting that the Facebook corpora have a greater prevalence of the categories 'Affiliation' and 'Achievement' compared to the literary and Lepkyi's corpora. Both Lepkyi and social media users tend to focus on current events, as shown by the higher percentage of present-time markers in the LIWC's 'Focuspresent' category.

Social media narratives have fewer cognitive processing markers compared to Lepkyi's and literary texts, while media stories contain more long words. According to Tausczik and Pennebaker (2010, p. 35), the use of words indicating cognitive mechanisms and those with more than six letters (Sixltr) are signs of complex language resulting from analytical thinking. Longer words are negatively correlated with their frequency in the text and positively correlated with the lexical variety and density quotient of the text (Kushnir et al., 2016).

When describing a past event, the use of causal words (such as 'because' and 'effect') and insight words (such as 'think' and 'consider') can suggest the active process of reappraisal. A reanalysis of six studies on expressive writing by Pennebaker et al. (1997) revealed that an increase in the use of causal and insight words by authors led to greater improvements in their health. This study suggests that actively processing an event while engaging in emotional writing can lead to better health outcomes when coping with traumatic stress. The frequency of causal and insight words has increased, which is comparable to the use of reconstrual statements. Previous studies have shown that discussing and reconstructing traumatic events leads to better health outcomes after trauma (Boals & Klein, 2005).

The frequent use of first-person plural pronouns can have different implications depending on the context in which they are used. For example, when 'we' is used to promote interdependence, such as in the phrase 'we can do this,' it may enhance group cohesion. However, if used to assign tasks indirectly, it may lead to resentment (Pennebaker & Tausczik, 2010). Therefore, it is unlikely that the latter connection of assigning tasks can be identified in War Narratives.

Simmons et al. (2008) conducted an investigation which found that an increase in the use of the second-person pronoun 'You' had a negative effect on the quality of the relationship, despite it being hypothetically a positive indicator. The study revealed that using such pronouns indicated a sense of hostility and a willingness to confront the interlocutor.

The Facebook narratives display the lowest 'You' ratio, suggesting warmer relationships. The data is consistent with our research findings, as 'You' and 'Anger' have the lowest frequencies in War Testimonies posted on Facebook compared to other corpora.

The study also analysed the emotional content of the narratives by identifying words that conveyed positive and negative emotions. While none of the generic LIWC categories showed a significant difference across the corpora, the 'Anger' category was found to be the most underrepresented in Facebook stories. When individuals feel uncertain or insecure about their topic, they tend to use polite language and filler words such as 'well', 'I mean', and 'you know'. Lepky's texts were less formal (0.97) and saturated with swear words (0.5). Both Lepky and the authors on Facebook expressed certainty and confidence in the topics they were addressing.

Categorical – Dynamic Indices	Lepky's war narratives (n=31)	Facebook corpus (n=354)	Literary corpus (n=100)	F	p
Categorical	5.87 (.92)	5.76 (6.39)	5.35 (.48)	.230	.795
Dynamic	4.23 (.72)	3.39 (2.35)	3.85 (.63)	3.906	.021*

Table 2: Mean Percents (Standard Deviation in Parenthesis) for Categorical and Dynamic Indices, by Corpus; p < .05.

The Dynamic Index of 'Writings from the War' on Facebook was found to be the lowest at 3.39. In contrast, Lepky's war narratives had a significantly higher Dynamic Index of 4.23 compared to two other corpora. However, further clarification was needed to understand why Lepkyi had the highest rates of both Categorical and Dynamic Indices, as it is somewhat ambiguous (see Table 2). To this end, we used an alternate and consistent approach suggested by Pennebaker et al. (2014). Thus, Categorical and Dynamic Index (CDI) was calculated integrally following formula (3) where a higher CDI score is associated with a writing style that uses more nouns, adjectives, and prepositions, and fewer pronouns, verbs, adverbs, and conjunctions:

30+adjective+preposition personal pronoun-impersonal pronoun-conjunctive-adverb-negation (3)

The results confirm the same tendency: the Index in Lepky Corpus is the lowest (20.56) whereas in Facebook it is the highest (24.73).

The summed figures of the Categorical and Dynamic Index suggest that Facebook stories have a more categorical style compared to Lepkyi's works, which have a more 'holistic' style (Nisbert et al., 2001) of the event description. This finding confirms our previous research based on Facebook posts by war witnesses during the first six months of the conflict. The trend observed during that period continued throughout the year.

Discussion and Conclusion

Regarding Research Question 1, the most outstanding finding from the results previously reported by Zasiekin et al. (2022) is that the LIWC category 'Achievement', which was insignificant at that time, has gained significance in the current Facebook narratives. As this category, according to McClelland (1961), reflects one of the basic human needs, its salience in the narratives discovered by LIWC suggests that Facebook users are praising their progress after a year of resistance and fighting against the enemy. Overall, the findings suggest a greater reliance on social ties at the personal, family and societal levels during wartime than in peacetime, as the social media narratives revealed three prominent LIWC categories related to social relationships: 'We', 'Social', and 'Family'. Thus, the predominant role of 'Affiliation' and 'Achievement' in the Facebook narratives is consistent with the more substantial impact of the 'We', 'Family' and 'Social' categories, which amplify the effect of 'togetherness'. This finding could indicate that collectively, the people facing disaster settings share values of trust, support, and protection. The authors also express a desire to unite in response to the threat.

In contrast, Lepky's stories showed insignificance in the aforementioned categories, but highlighted the psychological category of negative emotions, specifically 'Anger', along with cause and cognitive processes.

Research Question 2 in our study aimed to define the style of presentation of the war experience, whether analytical or narrative. The study revealed a significant deviation in the Dynamic Index across the three corpora, with the highest dynamic index found in Lepky's narratives (see Table 2). In contrast, the war testimonies on social media demonstrate just an opposite tendency towards analytical thinking. It confirms the presence of collective resilience features in Ukrainians as the ability to cope mentally and emotionally with a crisis due to their expression of categorising and conceptualising the events described in their narratives.

The narrative style in Lepky's texts and their low saturation with social markers and drives, but the considerable intensity of 'Anger', suggest that he may have been experiencing emotional distress, if not post-traumatic stress disorder (PTSD) (see Zasiekina et al., 2023), as a result of the suffering of the war and its aftermath at the time of writing. His stories exhibit a more narrative style and are sparsely saturated with social markers and drives such as 'We', 'Family', 'Affiliation', and 'Achievement'). As this study does not experimentally test for PTSD, we leave it out of the scope of this study.

Study Limitations and Future Directions

This study has several limitations that need to be addressed. Due to its cross-sectional design, it cannot draw any conclusions regarding causality. Additionally, pre-war Facebook linguistic data from users was not available. One potential direction for further research is to compare Lepky's war narratives with the literary prose of other Ukrainian writers on the ongoing Russia-Ukraine war. It would be fruitful to compare the discussed corpora of narratives with the "Narratives of War" (NoW) corpus compiled by researchers from McMaster University in Canada and Lesya Ukrainka Volyn National University in Ukraine as the result of joint project between the institutions. Further research is required to better understand the psycholinguistic features of war narratives and to assess the longer-term effects of war on the language and mental health of civilians. Moreover, it would be beneficial to include a new section of testimonies posted on Facebook by Ukrainians, which would provide psychological and linguistic insights into the experiences of individuals two years after the onset of the war.

Acknowledgements

The research was supported by a grant from the British Academy.

References

Aydin, Ciano. "How to Forget the Unforgettable? On Collective Trauma, Cultural Identity, and Mnemotechnologies." In: Identity 17/3 (2017), pp. 125–37. Accessed 3 December 2024 https://doi.or/10.1080/15283488.2017.1340160

Biber, Douglas. Variation across speech and writing. Cambridge University Press, 1988.

Bilyk, Nadiia. "Bohdan Lepky's contribution to strengthening Ukraine's integration to the global cultural process." In: Ukraine–Europe–World, L. Aleksievets, Ed. Ternopil: Ternopil Hnatiuk National Pedagogical University 1 (2008), pp. 108-118. (in Ukrainian)

Boals, Adriel, & Klein, Kitty. "Word Use in Emotional Narratives about Failed Romantic Relationships and Subsequent Mental Health." In: Journal of Language and Social Psychology 24/3 (2005), pp. 252–268. Accessed 3 December 2024 https://doi.org/10.1177/0261927X05278386

Date, Saroj, Sachin Deshmukh, Ryan Boyd, Ashwini Ashokkumar, & James W. Pennebaker. "Designing of a Novel Framework for Marathi Natural Language Processing: MR-LIWC2015". In International Journal of Intelligent Systems and Applications in Engineering 12/11s (2024), pp. 1-14. Accessed 3 December 2024 https://ijisae.org/index.php/IJISAE/article/view/4414.

Eisler, David F. Writing Wars: Authorship and American War Fiction, WWI to Present. University of Iowa Press, 2022. Accessed 3 December 2024 https://doi.org/10.2307/j.ctv2x00w38

Isaiuk, Olesia. The Russians of Bohdan Lepky: An unsuccessful game in Europe. In: Zbruc August 19, (2022). Accessed 3 December 2024 https://zbruc.eu/node/112872 (in Ukrainian)

Jordan, Kayla N., & Pennebaker, James. "The exception or the rule: Using words to assess analytic thinking, Donald Trump, and the American presidency." In: Translational Issues in Psychological Science, 3/3 (2017), pp. 312-316. Accessed 3 December 2024 https://doi.org/10.1037/tps0000125

Karalis, Magdalene. "The Information War: Russia-Ukraine Conflict Through the Eyes of Social Media". In: Georgetown Journal of International Affairs. February 2, 2024. Accessed 3 December 2024 (https://gjia.georgetown.edu/2024/02/02/russia-ukraine-through-the-eyes-of-social-media/)

Krylova-Grek, Yuliya. "Psycholinguistic Approach to the Analysis of Manipulative and Indirect Hate Speech in Media". In: East European Journal of Psycholinguistics 9/2 (2022), pp. 82-97. Accessed 3 December 2024 https://doi.org/10.29038/eejpl.2022.9.2.kry.

Kushnir, Oleg, Bryk, Oleg, Dzikovskyi, Viktor, Ivanitskyi, Lyubomyr, Katerynchuk, Ivan, & Kis, Yaroslav. "Statistical distribution and fluctuations of sentence length in Ukrainian, Russian, and English corpora". In: Bulletin of Lviv Polytechnic National University, Information Systems and Networks 854/1 (2016), pp. 228–239 (in Ukrainian).

Lanovyk, Zoriana, Lanovyk, Mariana. "'For war is war ...'": Bohdan Lepky as a writer of the "lost generation". In: "The Wind of Native Podillia Swang My Cradle": A Creative Phenomenon of Bohdan Lepky. In: Book of Abstracts of International research Conference dedicated to 150th birth anniversary of the writer. Ternopil: Ternopil Hnatiuk National Pedagogical University, (2022), pp. 25-29. (in Ukrainian)

McClelland, David. Achieving Society, 92051. Simon and Schuster, 1961.

Nisbett, Richard, Peng, Kaiping, Choi, Incheol, & Norenzayan, Ara. "Culture and systems of thought: holistic versus analytic cognition." In: Psychological Review, 108/2 (2001), pp. 291–310. Accessed 3 December 2024 https://doi.org/10.1037/0033-295X.108.2.291

Pennebaker James, Beall Klihr Sandra. "Confronting a traumatic event: Toward an understanding of inhibition and disease." In: Journal of Abnormal Psychology 95/3 (1986), pp. 274–281. Accessed 3 December 2024 http://dx.doi.org/10.1037//0021-843X.95.3.274

Pennebaker, James, Boyd, Ryan, Jordan, Kayla, & Blackburn, Kate. The development and psychometric properties of LIWC2015. Austin, TX: University of Texas at Austin, 2015.

Pennebaker, James, Chung, Cindy, Frazee, Joey, Lavergne, Gary, Beaver, David. "When small words foretell academic success: the case of college admissions essays." In: PLOS ONE 9/12 (2014), pp. e115844.

Pennebaker, James, & Seagal, Janel. "Forming a Story: The Health Benefits of Narrative." In: Journal of Clinical Psychology 55/10 (1999), pp. 1243-1254.

Suproniuk, O. Book legacy of Bohdan Lepky in his struggle for creating a new cultural narrative of Ukraine. In: The Library Bulletin 4/268 (2022), pp. 3-20. Accessed 3 December 2024 https://doi.org/10.15407/bv2022.04.003 (in Ukrainian)

Tausczik Yla, Pennebaker James. The psychological meaning of words: LIWC and computerized text analysis methods. In: Journal of Language and Social Psychology, 29/1 (2010), pp. 24-54. Accessed 3 December 2024 https://psycnet.apa.org/doi/10.1177/0261927X09351676

Zasiekin, Serhii. Psycholinguistic Regularities of Reproducing Literary Texts in Translation (Based on the English and Ukrainian Languages). Unpublished DSc thesis. Karazin National University of Kharkiv, 2020. (in Ukrainian)

Zasiekin, Serhii, Kuperman, Victor, Hlova, Iryna, & Zasiekina, Larysa. War stories in social media: personal experience of Russia-Ukraine war. In: East European Journal of Psycholinguistics, 9(2) (2022). 160-170. Accessed 3 December 2024 https://doi.org/10.29038/eejpl.2022.9.2.zas

Zasiekina, Larysa, Zasiekin, Serhii & Kuperman, Victor. Post-traumatic stress disorder and moral injury among Ukrainian civilians during the ongoing War. In: Journal of Community Health, (2023), 1-9. https://doi.org/10.1007/s10900-023-01225-5

Zasiekina, Larysa, Leshem, Becky, Hordovska, Tetiana, Leshem, Neta, & Pat-Horenczyk, Ruth. "Forgotten Stories of Women: Intergenerational Transmission of Trauma of Holodomor and Holocaust Survivors' Offspring." In: East European Journal of Psycholinguistics 8/1 (2021), pp. 137-158. https://doi.org/10.29038/eejpl.2021.8.1.zas

Sources

Gallagher, Matt. "You Don't Have to Be a Veteran to Write about War," LitHub, February 2, 2016, Accessed 3 December 2024 https://lithub.com/you-dont-have-to-be-a-veteran-to-write-about-war/

Lepkyi, Bohdan. Vybrani tvory. 2 Volumes. Nadiia Bilyk, Nataliï Havdyda, (Eds.). 2nd ed. Smoloskyp, 2011.

Lepkyï, Bohdan. Pysannia. Ukr. Nakladnia. Writings from the War, 1920. Accessed 3 December 2024 https://www.facebook.com/WritingsFromTheWar

NoW: Narratives of War: Virtual Exhibit of Written Testimonies of the Russia-Ukraine War. Accessed 3 December 2024 https://now.omeka.net

Zasiekin, Serhii. Literary texts Ukrainian 100. Mendeley Data, V1, 2022. Accessed 3 December 2024 https://doi.org/10.17632/9brrpc8zy8.1

Strategic Deception
Russian Trolls' Attempts to Reframe the Annexation of Crimea for English-speaking Audiences on Twitter/X

Maksim Markelov

Abstract

A growing body of research on state-sponsored social media actors has raised concerns over attempts by hostile states to influence public opinion online. The annexation of Crimea is one such high-profile cases, in which Russia was accused of using media technologies to influence Western publics. I use Twitter's/X's and Clemson University's Anglophone data on Internet Research Agency (IRA) and GRU (Russian military intelligence agency) actors and compare it with other users' ("non-troll") tweets on the topic of the annexation to understand how the linguistic and discursive practices employed by the state-sponsored actors identified by Twitter/X as "Russian trolls" evolve with time and context. I use a mixed methods methodological framework, which includes quantitative computational instruments for data collection and processing, qualitative Fairclough's (2001) Critical Discourse Analysis, tools of Systemic Functional Linguistics, Bakhtin's dialogic analysis, and collocation analysis to elucidate two crucial under-researched aspects of the annexation: (a) the annexation as a pivotal moment that reshaped Russia's social media landscape and (b) Russia's turbulent societal and social media environment after the annexation, which created fertile ground for language change. I argue that that most trolls' tweets align with the Kremlin's master narratives about the annexation, however, there are some significant outliers, which are thoroughly examined. The use of language observed is strategic, and the changes in language use indicate important differences between "troll" and "non-troll" messages.

Keywords

Twitter; trolling; computational methods; data processing; dialogic analysis; master narratives

Key definitions

Below are the definitions of several terms adopted to inform my analysis. The terms are heavily contested owing to their tendency to serve both as academic categories, and tools of practice used by social and political actors. However, as Melvin Richter points out in his seminal work "The History of Political and Social Concepts", such terms remain useful precisely due to their ambiguity (Richter 1995: 21). Crucially, identifying mis/disinformation or propaganda falls beyond the scope of this study, which primarily focuses on language use of online actors identified as "Russian trolls" by Twitter/X in the context of the annexation of Crimea.

Trolling – "the deliberate (perceived) use of impoliteness [and/or] aggression, [and/or] deception and/or manipulation in [computer mediated communication] to create a context conducive to triggering or antagonising conflict". (Hardaker 2013: 79).

Propaganda – "communication designed to manipulate a target population by affecting its beliefs, attitudes, or preferences in order to obtain behaviour compliant with political goals of the propagandist." (Benkler et al. 2018: 29).

Disinformation – fake or inaccurate information that is intentionally spread to mislead and/or deceive (Shu et al. 2020: 2).

Misinformation – "inaccurate information [that] can mislead people whether it results from an honest mistake, negligence [or] unconscious bias, [without the intent to communicate false information on purpose]." (cf. Fallis 2015; Benkler et al. 2018).

In July 2023, Twitter was rebranded to X. This work refers to it as Twitter since the data was collected under its former name. All examples of tweets in this article are unchanged from the originals.

Introduction

More than a decade ago, Morozov (2011) argued that the perception of social media as a fully open and liberal space with significant democratising potential had largely waned, and in many cases, states had managed to co-opt this technology and individual online personalities (e.g., bloggers, opinion-makers) to use them for their political ends. In 2024, the accumulating evidence of such activities leaves little doubt about the extent of social media platforms' use for political manipulation and control. Social media landscape has changed dramatically since Morozov's initial observations as its role in elections, the formation of public opinion, and even in inciting social unrest has been scrutinized, revealing a complex web of influence that extends far beyond simple user interactions.

This issue, however, does not lend itself to reductive binary interpretations, whereby hostile states' actors are seen in clear opposition to "ordinary" social media users. This is due to a general coarsening of public discourse and the widespread use of derogatory terms against political opponents by both state and non-state actors, both in neo-authoritarian regimes and Western democracies (e.g., the UK during and post-Brexit (cf. Douglas 2021), the U.S. in the lead-up to, during and after Trump's presidency (cf. Paul 2021)). Still, in contemporary Russia, the language of intolerance is often taken to extremes by state-controlled actors with widespread use of derogatory terms in state-controlled media, frequent aggression in Russian political YouTube (Bodrunova et al. 2021) and on social media (Pronoza et al. 2021). The present analysis focuses on one such case, detailing an attempt by Russian state-sponsored "trolls" at reframing the 2014 annexation of Crimea.

Extant research on the topic affirms that hate language, among other linguistic tactics, can be employed intentionally to influence both domestic and foreign audiences on social media platforms (e.g. before and/or during elections), while state-sponsored "trolling" can be regarded as an instrument of geopolitical influence (e.g. in the case of Russia, China, and Iran) (e.g. Glenski et al. 2019; Ghanem et al. 2019). Technological factors such as wide Internet/social media accessibility worldwide (5.44 billion Internet and 5.07 billion social media users as of April 2024 respectively) ("Number of internet" 2024), as well as innovations in automation and machine learning technologies make it easier than ever before for such actors to organize and run operations of greater scope, on a greater scale, and with more precision (cf. e.g. Bradshaw & Howard 2018).

I argue that the annexation of Crimea was a pivotal moment in history that reshaped Russia's social media landscape and marked the transition from sporadic domestic social media state trolling efforts to elaborate organised trolling campaigns, both domestic and foreign. Further, drawing on Milroy & Milroy (1985), Traugott & Dasher (2001), who posit that language change often occurs during periods of turbulence and upheaval in society, I propose that the economic and societal turbulence in post-annexation Russia created fertile ground for language change. Examining Russian Anglophone trolls' tweets in this context is important because it allows to reveal strategic foci and specific tactics and of state-sponsored social media actors, such as IRA, when targeting a much larger global user base (compared to Russophone audiences), in a different linguacultural context.

Crucially, I contend that Russian state-sponsored social media actors' strategies involve the reshaping of the very tools with which narratives are constructed and understood, including language, cultural and historical contexts.

The "troll factory"

Russia's best-known state-sponsored "troll factory", the Internet Research Agency (IRA), began its operation in 2009, targeting first a domestic Russophone audience in Russia during Russia's regional elections and, by 2012-2013, international audiences, including US voters on Twitter, posting content in a variety of languages (cf. "Расследование РБК" 2023, Bradshaw & Howard 2018). After the annexation of Crimea, IRA trolling activity on Twitter and other social networks began to grow rapidly, more than tripling on Twitter alone by late 2014-early 2015 (Howard et al 2018). The unprecedented volume of posts specifically targeting overseas audiences in that period compared to the previous years, suggests this IRA's campaign constitutes its first major attempt at influencing a foreign audience on social media.

Yevgeny Prigozhin, the founder of the "troll factory" and former leader of the Wagner Group (Russian state-funded paramilitary company), along with Yevgeny Zubarev, the head of "Federal News Agency" (one of the largest media outlets in Prigozhin's extensive propaganda network), later publicly confirmed scholarly findings into "the factory's" trolling efforts including its interference with the 2016 US Presidential Election ("Wagner chief admits" 2023). Following Prigozhin's rebellion collapse in June 2023 and his subsequent death in an August 2023 plane crash, his media empire, including the "troll factory," was allegedly shut down ("Расследование РБК" 2023). However, its trolling operations on social media seem to have persisted uninterrupted ("Как работали СМИ" 2023). Reports indicate that Yevgeny Prigozhin's son, Pavel, may assume leadership, collaborating with former manager Ilya Gorbunov to maintain and revive "the factory's" media assets ("Пригожин мертв" 2023).

The Annexation

In November 2013, mass protests, later known as "Euromaidan," erupted in Kyiv, Ukraine, following President Viktor Yanukovych's refusal to sign the EU Association Agreement. These protests turned violent by early 2014, leading to a crisis settlement agreement on 21 February 2014, mediated by the EU and Russia, involving constitutional changes, a unity government, and early elections. However, the deal collapsed when Yanukovych fled Ukraine after failing to resign, leading to the Ukrainian parliament forming a new government. Concurrently, unmarked soldiers (later identified as Russian) seized strategic locations in Crimea, culminating in a disputed referendum in March 2014, followed by the annexation of Crimea by Russia. This move, unrecognized internationally and condemned by Western nations, led to sanctions against Russia (cf. "Conflict in Ukraine" 2023).

Despite this, significant majorities in Crimea (82-83%) and Russia (around 88%) reportedly supported the annexation, with consistent polling data from 2014 to 2019 confirming this stance ("Крым" 2021, "To Russia With Love" 2020).

Data description and availability

The analyzed data comprises two text corpora: (1) Russian "Troll" corpus (IRA, GRU) with 97,201 case-relevant tweets in English, and (2) a Control ("non-troll") corpus with 454,227 case-relevant tweets in English. The "troll" corpus includes the best available Twitter Russian "troll" data widely used in academic research: Twitter's own datasets for IRA and GRU-linked actors ("Twitter Moderation" n.d.) and Clemson's Social Media Listening Center datasets, derived from Twitter's IRA and GRU account data ("fivethirtyeight/russian-troll-tweets" 2018) – both publicly available. The control corpus data are openly available in Harvard Dataverse at https://doi.org/10.7910/DVN/GDNVIZ. Due to ethical and legal constraints, only Tweet IDs of the "non-troll" tweets are published, allowing for retrieval from Twitter/X provided the tweets remain publicly visible.

The author has obtained ethical approval from the relevant ethical committee.

Methodology

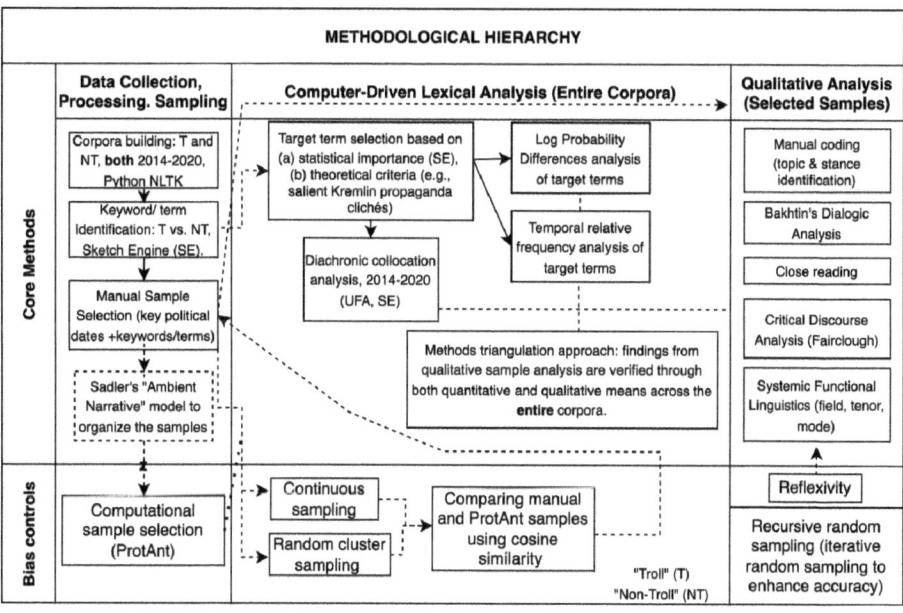

Figure 1: Methodological framework.

This study uses a mixed methods approach, consisting of seven stages executed in the following order as each stage builds on the previous one (Fig.1). Each method provides distinct strengths: computational data collection and pre-processing ensure a robust dataset, keyword identification highlights main thematic foci, and sampling methods control for bias. Further, the qualitative approaches offer insights into the strategic use of language and its socio-political contextualisation. Using these methods together allows for a comprehensive examination of both broader quantitative patterns and qualitative nuances in troll and non-troll discourses.

1. Data collection.
I focus on the period from January 1, 2014 to December 31, 2020. Hence, only the tweets published in this period (both "troll" and "non-troll") are included in the analyzed corpora.

To ensure that the "non-troll" corpus does not contain any "troll" tweets, I verify that all "troll" accounts are suspended, and their tweets are unavailable for collection using tweet and user IDs. Then, the tweets for the "non-troll" corpus (2014-2020) are collected using Twitter Academic Research program's Application programming interface (API) ("Step-by-step" n.d.). The following case study search terms are used: "crimearussia, crimeaukraine, crimeaannexation, crimea2014, crimeaputin, crimea russia, crimea ukraine, crimea annexation, crimea 2014, crimea Putin, Russia, Ukraine, news crimea, Russian, Ukranian [misspelling is intentional], politics." The daily tweet limit is set to 500 (up to 31 tweets per keyword), to match the volume of the unfiltered "troll" corpus and ensure a bona fide sample of tweets distributed as evenly as possible across time and topics. This limit helps prevent overrepresentation of tweets from specific days and the "burstiness" (sporadic increase in frequency) of search terms (Altmann et al. 2009).

2. Data pre-processing (filtering, normalization, tokenization) using Python Natural Language Toolkit processing algorithm.
For the "troll" corpus pre-processing comprises: (1) filtering by (a) year (2014-2020); (b) tweet language (English); (c) content type (not retweets). (2) Given the case study's focus, all tweets in the "troll" corpus lacking at least one of the search terms used for collecting the "non-troll" tweets are filtered out resulting in 97,201 case-relevant tweets (out of over 2.4 million total tweets in English).

For keyword/term identification, both "troll" and "non-troll" corpora are (a) normalized (removing hashtag symbols (#), punctuation, numbers, and hyperlinks to make the texts more easily readable by the machine), (b) tokenized (split into constituent parts – words), (c) Part of Speech tagged.

For the qualitative part of the analysis, the original corpora before normalization/tokenization are used owing to punctuation, emoticons, emojis, and hyperlinks being instrumental to the analysis as extralinguistic markers signalling, for example, verbal aggression.

3. *Keyword/terms identification in the corpora using Sketch Engine keyword tool.*
It identifies keywords/terms by dividing the relative frequency of a word/term in a target corpus by the relative frequency of the same item in a reference corpus, applying measures to reduce the effects of sporadic bursts in distribution (cf. Kilgarrif 2015). Using default parameters, a list of words/terms that are distinctive of the "troll" corpus compared to "non-troll" corpus is generated.

4. *Sample selection based on (a) the dates of key political events before, during, and shortly after the annexation of Crimea (in "Conflict in Ukraine" 2023) and (b) 13 salient case-study-relevant key words/terms identified prior: "Crimea, Crimean, Rada, referendum, Russian envoy, ukrainianprotests, Ukrainian crisis, Ukrainian governments, Ukrainian official, Ukrainian people, Ukrainian president, Ukrainians, Ukranian [misspelling is intentional]." This resulted in 833 "troll" and 838 "non-troll" samples. Selected samples are exclusively utilized for the qualitative analysis, while quantitative analysis always encompasses the entirety of the corpora.*
To control for bias, I use: (1) continuous sampling (2014-2020), (2) random cluster sampling: selecting ten random tweets for each calendar quarter in the corpora, (3) ProtAnt software for comparative keyword frequency analysis and prototypical text identification (Anthony & Baker 2015), and (4) cosine similarity for comparing manual and ProtAnt samples, confirming high representativeness in "troll" (0.77) and "non-troll" (0.91) samples.

5. *Qualitative analysis of the samples using relevant elements of The Fairclough (2001) model of Critical Discourse Analysis (CDA), Systemic Functional Linguistics (SFL), and Bakhtin's dialogic analysis.*
Firstly, to address the fragmented nature of the data, I apply Sadler's (2021) "ambient" narrative model (a mode of collective storytelling on platforms like Twitter/X whereby many social media users contribute to larger loosely defined narratives, for instance, using hashtags). This model guides my analysis by considering: (a) fragment interpretation order as secondary to understanding the overall timeline of events, (b) inherent repetition in "ambient narratives", which can be used strategically for emphasis or audience engagement, (c) the intent for narrative fragments to be shareable, and (d) the elevated role of readers in piecing together narratives, making a dialogic approach to studying these multi-voiced narratives beneficial.

To that end, I draw on Bakhtin (2008) to analyze (a) performativity (what the tweet is "doing" rather than what it is "saying": provoking; mocking; unifying (with other "trolls"); disrupting, etc.) and (b) dialogism (which voices and points of view the text implicitly endorses, negates, engages with, responds to, etc.).

CDA complements the analysis and offers critical insights along the dimensions of text (tweet text itself), discourse practice (how it is produced and is intended to be received) and social practice (larger social context framing the discourse). It is applied together with field (describes the topic of discourse), tenor (identifies the participants and their relationships), and mode (indicates the manner and formality of communication) concepts of SFL.

6. Target word selection for subsequent collocation analysis.
I trace specific instances of use of the salient keywords identified at stage three as well as the terms that have acquired widespread use in media and social media discourses related to the annexation: "Ukranian [misspelling is intentional], Ukraine, Crimea, Annexation, Occupation, Protectorate, Protest, Referendum, Misinformation, Invasion, War, return home, intervention, independence, re-unite, reunification, reunion, Crimea federal district, republic of Crimea, Crimea river, terror-Russia, Russian witch-hunt, Russian armed forces."

7. Diachronic collocation analysis of the selected target words using Usage Fluctuation Analysis (UFA) (McEnery et al. 2019) algorithm and the Sketch Engine software.
This stage of analysis is theoretically grounded in Traugott and Dasher's (2001) Invited Inferencing Theory of Semantic Change (IITSC), which comprehends the interrelation between lexical sematic change and pragmatic factors. I trace collocational behaviour of the same target words in the "troll" and "non-troll" corpora. UFA tracks changes in words' use over time by examining their diachronic collocational behaviour. Based on the collocations that occur in the vicinity of a target word through time, the algorithm identifies patterns, and subsequently the periods when a word's usage might be changing or remaining stable (McEnery et al. 2019). This stage of the analysis focuses on examining specific themes and patterns by tracing a word/term's use diachronically through collocation analysis.

Finally, to determine the prevalence of selected terms, their Log Probability Differences between the "troll" and "non-troll" corpora are calculated. Log Probability Difference measures the likelihood of terms appearing in one corpus over another based on their actual frequency, ensuring normalization for differences in corpus size and enhancing rigor and formalization.

Dominant themes

To identify the dominant themes, I analysed the selected samples (methodology, step four) categorizing them by topic. This involved a close reading and manual coding by the author, with particularly prominent themes identified as predominant (based on aggregated list of all identified themes and tweet numbers per theme). In the "troll" samples, the most dominant theme is "diminishing the military potential of Ukraine and boosting the military potential of Russia", whereas in the "non-troll" set it is "condemning the annexation". Both corpora also contain the following less prominent themes: "drawing parallels between World War 2 and the Cold War and the Crimean crisis", "emphasizing the "inevitability" and "finality" of the annexation", "spreading implied threats about the potential use of nuclear weapons", "World War 3", "military action in other regions (in Eastern

Ukraine and the rest of the country, in Moldova (Transnistria))", "speculating about possible negative effects of the annexation for Russia", while, "suggesting that Crimea has always historically and rightfully been a part of Russia" is only present in "troll" sample set and is absent in the "non-troll" sample set.

"Trolls'" and "non-trolls'" language use

The language use patterns reveal a distinct difference between "troll" and "non-troll" perspectives on the annexation, with notable exceptions in both groups suggesting tactical use – while the "trolls" employ the "playing both sides" tactic, potentially aiming to draw attention and sharpen existing divisions, "non-troll" users prefer satire or sarcasm.

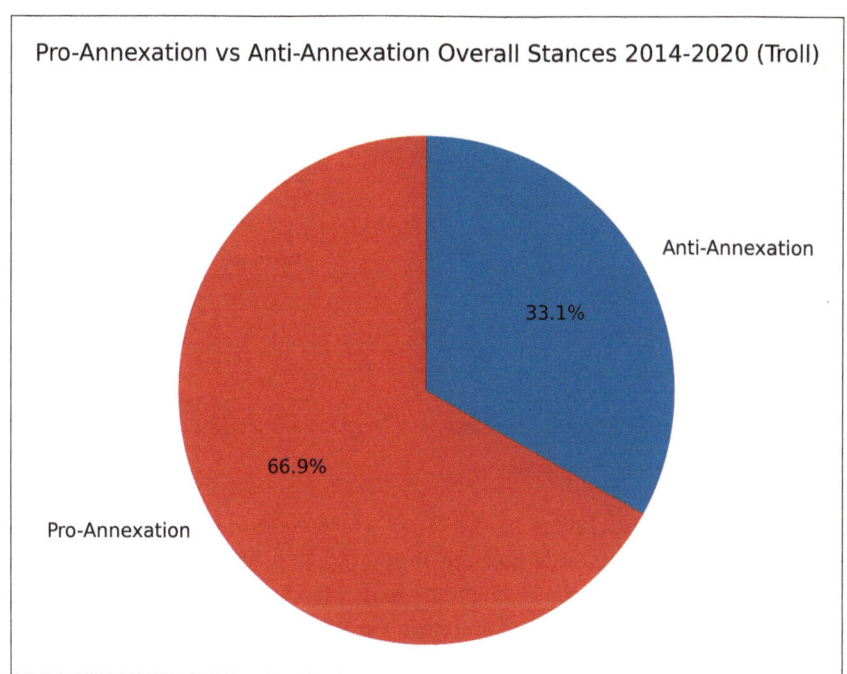

Figure 2. Ratio of Pro/Anti-Annexation Overall Troll Stances 2014-2020.

Most "troll" tweets (Fig.1) that characterize Crimea's union with Russia aim to present it as a legal, just, and anticipated event. The wording choices include "Pro-Russian separatism", "protesters' clash", "region's divide", "Russian intervention", "Intervening [in Crimea]", "Russian protectorate [over Crimea]", "[Crimea's] independence", "[Crimea] votes to join Russia", "Crimeans vote to re-unite with Russia", "[Crimea's] reunification with Russia", "Crimea's 'return home'", "Russia regained 'historic roots' in Crimea", "Crimea [is] Russia", "Crimea is Russian territory", "the reunion of Crimea and Russia", "Freedom for Crimea", "retake Crimea", etc. The "trolls" maintained an overwhelming pro-annexation stance throughout 2014, which gradually shifts to negative stance by 2016 often in a form of criticism of President Obama's handling of the Crimean crisis in the 2016 U.S. Presidential Election context.

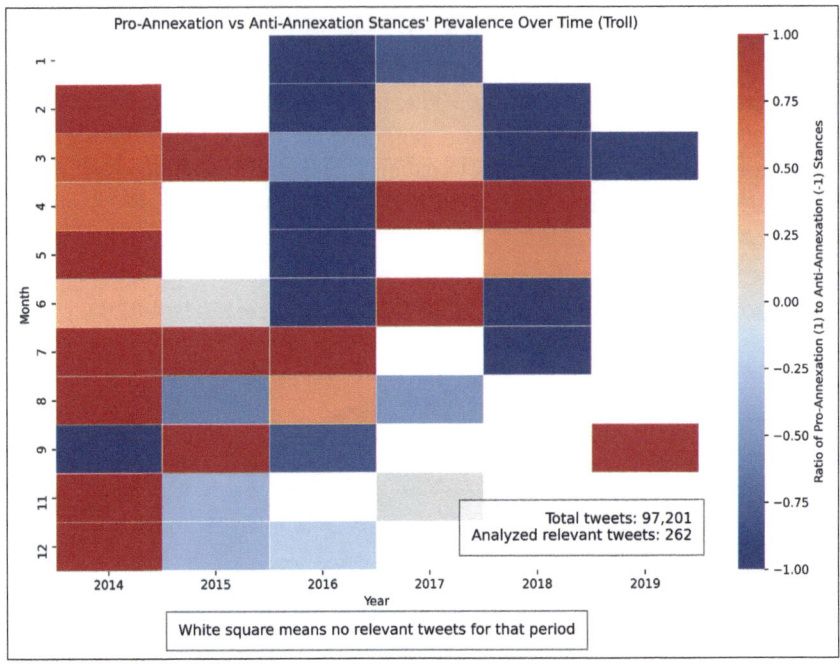

Figure 3. Pro/Anti-Annexation Troll Stances' Prevalence Over Time.

Identified vocabulary choices in the "troll" corpus are ideologically motivated and are in clear opposition to how the annexation is predominantly regarded in the "non-troll" corpus. Initially (February-early March 2014), Russia's actions in Crimea are presented as peacekeeping efforts in a violence-torn region. The "trolls" aim to portray Crimea as a socially and politically turbulent region with apparent divisions regarding its pro-Russian or pro-Ukrainian orientation: "Pro-Russian separatism", "protesters' clash" and "region's divide" particularly highlight this. The expressions "Pro-Russian separatism" and "region's divide" are only used in relation to Crimea, whereas "protests", "protestors" and "separatism" – in relation to Ukraine in general. This has significant implications: (1) before the referendum on the status of Crimea divisions in the region are highlighted, whereas after the Referendum – Crimea's pro-Russian orientation is emphasized, which aims to portray the annexation as Crimeans' choice, (2) "separatism" in Ukraine in general and "Pro-Russian separatism" in Crimea are used to convey two distinct ideas: the former – of unlawful activity, the latter – of Crimeans' purported desire to join Russia. Amid the region's purported divisive socio-political climate and strong pro-Russian sentiments, the "trolls" depict Russia's actions in Crimea as an "intervention" [as opposed to what it really is – "invasion"], which implies that such an "intervention" might be "necessary" or "justified" and is "aimed at improving the situation", rather than meddling in an independent state's internal affairs.

Later, following the Crimean Autonomous Assembly's declaration of independence (mid-March 2014), the main thematic focus shifts, and words such as "independence" and "protectorate" start appearing in the "troll" tweets, wherein Crimea is described as both (a) an autonomous region that can decide its own fate and (b) a region that needs Russian protection and support. Although not mutually exclusive, the suggestion of "protectorate" implies a forceful Russia's influence in Crimea, contrasting with, for instance, financial or other aid to support Crimea's independence. Further, "protectorate" is used in a positive sense, suggesting that the "trolls" might not have fully comprehended the larger historical context and baggage related to the system of imperial rule associated with the term. Conversely, they might have likened "protectorate" to "protection", believing that the former term is derived from the latter and is inherently positive.

The "troll" set:
(T1) "The ideal scenario would be if #Russia and the EU could agree on an informal protectorate over #Ukraine http://t.co/9BAOwoQzHT"".
(T2) "Occupying #Crimea as a Russian protectorate may be enough to prevent integration of Ukraine with the EU and NATO. http://t.co/zJieoSM3Tw".

Days before and during the Crimean "referendum" on joining Russia, this union is presented as legal and as a reflection of the Crimean peoples' will. The "trolls" speak of "voting" and "referendum" in Crimea and its willingness to "reunite" with Russia, aiming to superimpose the guise of democracy onto a process forced by significant Russia's military presence in the region and deemed illegitimate by the international community.

Finally, the "trolls" predominantly refer to the annexation as "reunification", "return home", "reunion", as opposed to "annexation", "seizure" or "occupation" in the "non-troll" set, echoing the most oft repeated clichés on Russian state television. The consistent "troll" lead in using, for instance, "return home" and "non-troll" lead in using "annexation" is evident (Fig.4,5) with several outliers discussed further.

Specifically, Crimea's historical roots are emphasized. The trolls implicitly refer to the fact that Crimea was a part of Russia historically and was separated from the Russian Soviet Federative Socialist Republic and transferred in 1954 to Ukrainian Soviet Socialist Republic on a whim of the Soviet government. The prefix "re" in many of the terms used by the "trolls", such as "reunification" or "regaining roots" is used to additionally stress that Crimea is "going back" to Russia, such as a territory that was once under another nation's temporary control and now returns to the "motherland".

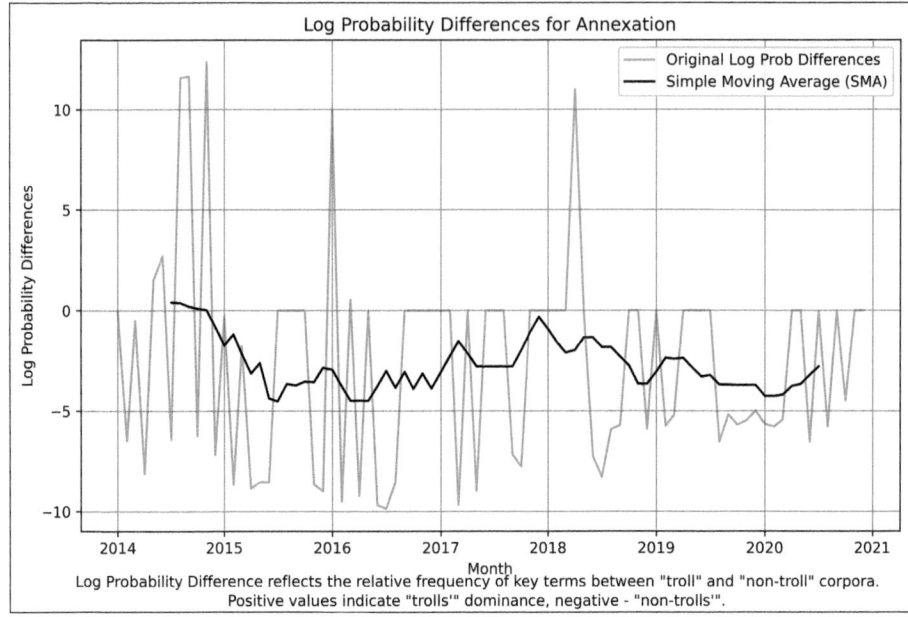

Figure 4. The use of "annexation" in "troll" vs. "non-troll" corpora (2014-2020).

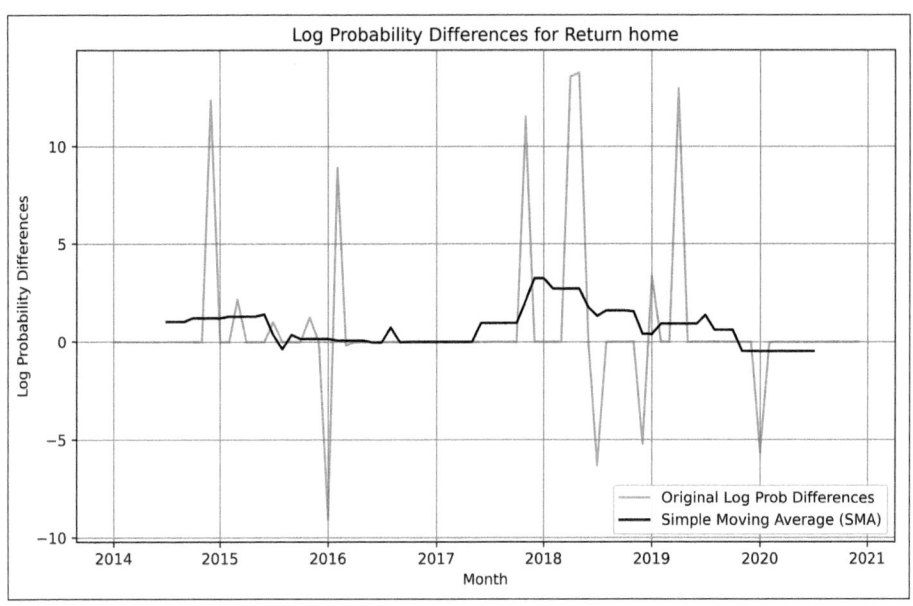

Figure 5. The use of "return home" in "troll" vs. "non-troll" corpora (2014-2020).

Covert deception

Other studies confirm that such "troll" behaviour, wherein propaganda from mainstream state media is amplified on social networks is to be expected (e.g., Keller et al., 2020). However, I have observed several posts that do not fit within that category and do not align with official Russian state media narrative. In those tweets, the words "annexation" and "occupation" are used, but "occupation" is presented as something positive and inevitable, and "annexation's" legal definition is contested. This tactic might be a part of an overall trolling strategy to "play both sides": to polarize and confuse the readers while exacerbating existing divisions. Benkler (2018) points out that this tactic was employed during the 2016 U.S. Presidential Election by media actors and is not exclusive to Russian state "trolls". This is a likely attempt by the trolls to reach the English-speaking target audience more easily using lexicon more familiar to them (i.e., "annexation" and "occupation") rather than translated versions of Russian mainstream media terms such as "reunification".

Using the "other side's" vocabulary and subverting its meaning appears to be consistent with the activity of state trolls in their efforts to sharpen divisions and maximize antagonism (cf. e.g., Benkler et al. (2018), Suk et al. (2022), Rains et al. (2023)). There is a possibility that it is a (a) deliberate bold antagonistic provocation, intended to be seen as such, or (b) an attempt at covert deception based on subverting the meaning of the other's term. Drawing on Bakhtin (2008), I apply a dialogic approach to better understand the nature of such subversion.

The "troll" set:
(T3) "#OSCE refused to consider #Crimea annexation as violation of #Ukraine's territorial integrity http://t.co/tRQlYKrx8s".
(T4) "@BBCWorld Annexation is the FORCIBLE acquisition of a state's territory by another state. What about referendum on the status of Crimea?".

Discourse in general and Twitter/X discourse in particular is dialogically oriented and cannot be viewed in isolation from other discourses (cf. Brock 2018). The above two tweets are situated within and are in dialogic relations with legal, military, democratic and other discourses. The legal discourse encourages adherence to the rule of law, internationally established conventions, national sovereignty, and core principles aimed at ensuring international peace and security. The military and democratic discourses in this case intersect with the legal one and do not promote the idea of using state armed forces outside of its internationally recognised borders without proper justification, whereas the idea of referendum as free expression of the peoples will is upheld in the democratic discourse. The "troll" responses challenge these values by justifying the blatant act of annexation and claiming it does not violate any laws, internationally accepted legal, military, or democratic principles.

According to Bakhtin (2008), in dialogue there is a continuous "tension" between "self" and "other", as well as between socially accepted discourses and discourses that one finds internally persuasive and can incorporate into one's own worldview. Through this dynamic process one can orient oneself in larger society and formulate one's opinions in the process of communication, through dialogue.

In (T3) and (T4), the "trolls" seek to engage in dialogue with the wider English-speaking Twitter audience under the identity of a person who believes that the annexation was a rightful and legal act. That political identity regards opposing views on the situation as invalid. The intended audience for these tweets may or may not be familiar with the official Western narratives on the annexation. However, they must recognize key terms like "annexation," "territorial integrity," and "referendum," and be politically literate enough to understand the context and implications of the tweets.

Suk et al. (2022) posit that Russian "trolls" have a strategic motivation to deceive their target audience. To achieve this, their content needs to be persuasive, and, in many cases, emotionally charged. The researchers demonstrated that greater syntactic complexity and the perception of posts as written by authentic U.S. users tended to lead to more retweets (ibid: 607). Further, posts with a negative sentiment are expected to be retweeted significantly more compared with positive ones or ones without negative sentiment (ibid: 605). Provoking others emotionally is also characteristic of typical troll behaviour: it can be a means to an end or an end in itself (cf. Hardaker 2010).

In (T3) and (T4), there are three aspects indicating that adopting the other's vocabulary and subverting its meaning is an attempt at a covert deception rather than blatant provocation. Firstly, the trolls adopt the identity of an "I, who accepts the annexation of Crimea" in clear opposition to the Western mainstream "Other, who views the annexation of Crimea as an illegal and wrongful act, deserving international condemnation", thus casting doubt on the widely accepted interpretation of the events. Secondly, since the posts were made by state-sponsored "trolls", we can assume that there was a motivation to deceive, aggravate, polarise or confuse the audience. Third, to achieve the desired outcome the posts were complex in terms of general syntax and word choice. Drawing on Suk et al. (2022), syntactic complexity in Anglophone contexts involves complex structures, such as dependency length (with greater length indicating higher complexity) and structures enhancing cognitive processing, such as appositional modifiers, nested prepositions, and clausal complements. It is a facet of overall linguistic complexity, reflecting text detail and associated with cognitive demand. To illustrate, in (T3) the dependency length (the distance between linguistic "heads" (words central to the structure) and dependants – words that modify the "heads" – is eight, as there are eight links between the "heads" and dependants. In (T4), the dependency length is eleven indicating greater complexity.

Examples (T3) and (T4) express negative sentiment (lexically – words such as "refused", "violation", typographically and stylistically – the use of all capital letters in "FORCIBLE", using a negative assertion in a form of a rhetorical question "What about referendum on the status of Crimea?") and trying to imitate the "Other" to seem more convincing. Therefore, the posts were designed to be more persuasive, draw more attention and retweets, as well as challenge the mainstream Western view on the annexation. The likely objective was to instill doubt among Twitter users acquainted with mainstream Western political discourse yet receptive to alternative interpretations.

Discursive (Re)Framing

Whenever the other side's vocabulary is not adopted, the "trolls" rely on euphemisms to avoid negative values (using "reunion" and "return home" to avoid "annexation", "invasion" and "occupation", "to tighten hold" instead of "continue the invasion", "intervention" instead of "occupation", "protesters" instead of "separatists", "militia", or "civil resistance").

Conversely, to intentionally create negative values, dysphemisms, such as "civil war" instead of "protests", "censorship" instead of "restrictions", "poverty" and "homeless" instead of "hardship", "neo-Nazi regime" instead of "Ukrainian government" are used. Overwhelmingly, euphemisms and dysphemisms are used throughout the "troll" set in a way that perfectly aligns such narratives with the Kremlin's propaganda, which might indicate that such language use is strategic. Although similar euphemisms and dysphemisms are also used sporadically in the "non-troll" set, the ideological polarity between the two is apparent: the "trolls" predominantly follow the Kremlin's strategic narratives and avoid negative values in relation to Russia and its actions in Ukraine, while framing Ukraine, its government, and the "West" negatively (particularly through the negative representations of Ukraine's Western backers). Conversely, the "non-trolls" employ euphemisms and dysphemisms without shying from criticizing both Russia and Ukraine. Compare:

The "troll" set:
(T5) "US pushed the #Ukraine into civil war. The country already lost Crimea, and is de facto losing #Donetsk and #Luhansk. http://t.co/7n1jN9mvQH".

The "non-troll" set:
(T6) "I hope my co-workers don't loose sight of what's important during this ukranian civil war, timely and accurate marketing emails #notreally".

Redefining "Ukraine"

Finally, I will briefly focus on one illustrative case of diachronic meaning evolution of the word "Ukraine" identified by means of collocation analysis.

Ukraine's state history is intricate, marked by periods of autonomy and foreign rule, which saw Ukraine as a part of the Russian Empire and the Soviet Union, culminating in its independence and international recognition in 1991. Ukraine's strategic location between Russia and the European Union, viewed by Moscow as a NATO buffer, along with historical, cultural, and economic connections, and access to the Black Sea, have underscored its geopolitical significance. Throughout Putin's presidency, the Kremlin has aimed to retain Ukraine in its sphere of influence across ideological, cultural, political, and economic dimensions, while challenging Ukraine's sovereignty and national identity, undermining its statehood and portraying it as a failed state at the highest level of Russian political discourse. Such Kremlin's narratives have sought to redefine "Ukraine", questioning the legitimacy of its nationhood and independence internationally recognized since 1991 (e.g., Ukraine as only "three regions" ("Путин заявил" 2023).

The collocational behaviour reveals a stark contrast in the way trolls and non-trolls imbue meaning in "Ukraine" (Table 1). "Trolls'" discourse is characterized by a dynamic and sensational array of associations, often linking Ukraine with controversial and provocative terms such as "Chernobyl", "Nazis", "Syria", and "terrorists". This suggests a likely attempt to incite fear, confusion, and distrust, framing Ukraine within a narrative of crisis and instability.

The collocations "Syria", "Venezuela", and "Bolivia" reflect the "trolls'" blaming of the US or the "the ruling class and the empire of finance" for the crises in these countries, equating the US, the West and "capitalists" to "terrorists". This framing of "Ukraine" in these contexts and drawing parallels between it and these states sought to equate "Ukraine" with politically unstable states, portraying it as a victim of Western interference. Consequently, this collocational pattern in "trolls" discourse suggests a pejorative shift in the connotation of "Ukraine," depicting it as a politically unstable state lacking clear governance and adversely affected by Western "meddling". In contrast, "non-troll" discourse maintains a consistent focus on immediate geopolitical realities, predominantly and consistently associating "Ukraine" with "Russia", and terms like "crisis" and "aid". Thus, while the "trolls" exploit emotionally charged and divisive topics attempting to shape perceptions, "non-trolls" present a more coherent and less sensationalized account, reflecting actual geopolitical contexts.

Year	Collocates	
	"Troll"	"Non-troll"
2014	-	Russia, crisis, Russian.
2015	Chernobyl, crisis, danger, emergency.	Russia, stop.
2016	Dutch, Poroshenko, Russia.	Russia.
2017	Collusion, Dems, Nazis, Syria, Venezuela, terrorists.	Russia.
2018	Missile, Syria, US.	Russia.
2019	Agenda, Bolivia, Georgia.	Russia, aid.
2020	-	Russia, China.

Table 1. Collocations of "Ukraine" over time in "troll" and "non-troll" corpora.

Discussion and conclusion

The findings illuminate the elaborate linguistic strategies utilized by Russian state-sponsored trolls, which align with overarching Kremlin narratives. This dovetails into broader issues regarding the adaptation of authoritarian state propaganda to foreign linguacultural and geopolitical contexts. In this case, it pertains to the adaptation of themes and discourse tactics by Russian state-sponsored social media actors to the Anglophone context of Twitter/X on the topic of the annexation. It aligns with the research of Benkler et al. (2018) and Suk et al. (2022), which examine how authoritarian regimes, such as Russia's, tailor narratives and discursive practices to target Western audiences.

The strategic use of language by these actors to reframe and amplify certain narratives and themes in conflict situations reflects broader state propaganda strategies aimed at manipulating public perception and legitimizing political agendas. The findings also relate to research by Bradshaw & Howard (2018) elucidating how dis/misinformation campaigns systematically employ rhetorical and discursive techniques aiming to shape opinion and exacerbate societal divides.

In the context of Ukraine, the trolls' deployment of euphemisms and dysphemisms – such as labelling an invasion as a "reunification" and portraying Ukraine as a "neo-Nazi regime", a politically unstable state and a "victim" of Western influence – serves to undermine Ukraine's sovereignty and national identity at the level of discourse. These linguistic manipulations align with the tactics identified by Ghanem et al. (2019) and Bodrunova et al. (2021), where aggressive and deceptive language is employed to reinforce state propaganda.

At the same time, Russian state-sponsored social media trolling targeting foreign audiences should not be viewed merely as an extension of state propaganda through direct translation into a new language. Nor should it be seen as genuine ideologically motivated efforts by the "trolls" to help achieve the Kremlin's objectives. The "trolls'" alignment with the Kremlin's narratives, as shown by empirical data, clearly stems from the management of the "troll factory" that directs the desired agenda and the financial incentives given to the trolls as employees. Their ideological stance or genuine efforts cannot be assumed. Conversely, research into the functioning of the Russian "troll factory" suggests that incentives are primarily financial (cf., e.g., "Как работали СМИ" 2023).

The strategic choices of Russia's state communication, as observed in the state "trolls'" output, aim to normalize the Kremlin's agenda through consistent reframing and amplification of certain themes, including via linguistic means such as the use of euphemisms and dysphemisms, imbuing the opponent's terms with different meanings, adapting and contesting them dialogically. The "trolls'" exploitation of the Ukraine-West dynamic thus illuminates efforts to bolster Kremlin's own ideological stance and justify its geopolitical actions in Ukraine. This is evident in the portrayal of NATO as a threatening force and the West's support for Ukraine as "meddling" in "Russia's sphere of influence".

The temporal analysis reveals how Russian state-sponsored social media actors, such as the IRA, dynamically adjust their linguistic tactics in response to evolving geopolitical contexts. Initially, the portrayal of Russia's actions in Crimea as peacekeeping efforts amid regional violence shifted to emphasizing Crimea's autonomy and the need for Russian protection post- "referendum". This tactical adaptation aimed to reframe and legitimize Russian geopolitical actions, aligning with broader Russian state propaganda tactics.

Examining the collocates of "Ukraine" over time further illuminates the "trolls'" strategies. In 2014, "non-troll" discourse centered on "Russia", "crisis", and "Russian", while the "trolls" largely overlooked these associations. By 2015, the "trolls" began associating "Ukraine" with negative terms like "crisis", "danger", and "emergency", contrasting with the "non-troll" consistent focus on "Russia". This pattern continued, with the "trolls" using increasingly aggressive and emotionally charged terms like "collusion", "Nazis", and "terrorists" by 2017. In subsequent years, collocations like "missile", "Syria", "Venezuela", "Bolivia" dominated "troll" associations of "Ukraine", aiming to reinforce a narrative of global conflict and instability linked to Ukraine. Meanwhile, "non-troll" discourse remained consistent with a focus on "Russia", occasionally incorporating other global actors like "China" in 2020. This temporal analysis highlights the "trolls'" strategic manipulation of language in an attempt to influence perceptions of Ukraine, bolster Russian state propaganda narratives, and shift blame onto Western influences, feeding into the Ukraine-West and the broader Russia-West dynamic.

Future research could quantify the impact of such trolling strategies on public opinion internationally and explore the psychological and sociological drivers behind the traction of specific narratives, including their effects on group identity, to unveil the mechanisms of influence and persuasion in state-sponsored trolling.

References

Altmann, Eduardo G., Janet B. Pierrehumbert, Adilson E. Motter (2009): "Beyond word frequency: Bursts, lulls, and scaling in the temporal distributions of words." PLOS ONE 4/11:e7678 (https://doi.org/10.1371/journal.pone.0007678).

Anthony, Laurence, Paul Baker (2015): "ProtAnt: A tool for analysing the prototypicality of texts." International Journal of Corpus Linguistics 20/3, pp. 273-292.

Bakhtin, Mikhail (2008): The Dialogic Imagination: Four Essays. Austin:University of Texas Press.

Benkler, Yochai/Robert Faris/Hal Roberts (2018): Network Propaganda: Manipulation, Disinformation, and Radicalization in American Politics. Oxford University Press.

Bodrunova, Svetlana S., Anna Litvinenko, Ivan Blekanov, Dmitry Nepiyushchikh (2021): "Constructive aggression? Multiple roles of aggressive content in political discourse on Russian YouTube." Media and Communication 9, pp. 181-194.

Bradshaw, Samantha, Philip N. Howard (2018): "The global organization of social media disinformation campaigns." Journal of International Affairs 71/1.5, pp. 23-32.

Brock, André (2018): "Critical technocultural discourse analysis." New Media & Society 20/3, pp. 1012-1030.

"Conflict in Ukraine: A timeline (2014 – eve of 2022 invasion)", August 22, 2023 (https://researchbriefings.files.parliament.uk/documents/CBP-9476/CBP-9476.pdf).

Douglas, Fiona M. (2021): Political, Public and Media Discourses from Indyref to Brexit: The Divisive Language of Union, Cham:Palgrave Macmillan.

Fairclough, Norman (2001): Language and Power, 2nd ed. London: Routledge.

Fallis, Don (2015): "What is disinformation?" Library Trends 63/3, pp. 401-426.

"fivethirtyeight/russian-troll-tweets", August 27, 2018 (https://github.com/fivethirtyeight/russian-troll-tweets/).

Ghanem, Bilal, Davide Buscaldi, Paolo Rosso (2019): "TexTrolls: identifying Russian trolls on Twitter from a textual perspective." (arXiv preprint arXiv:1910.01340).

Glenski, Maria, Ellyn Ayton, Josh Mendoza, Svitlana Volkova (2019): "Multilingual multimodal digital deception detection and disinformation spread across social platforms." (arXiv preprint arXiv:1909.05838).

Hardaker, Claire (2010): "Trolling in asynchronous computer-mediated communication: From user discussions to academic definitions.", Journal of Politeness Research 6/2, pp. 215-242.

Hardaker, Claire (2013): "'Uh.... not to be nitpicky, but... the past tense of drag is dragged, not drug.': An overview of trolling strategies." Journal of Language Aggression and Conflict 1/1, pp. 58-86.

Howard, Philip N., Ganesh, Bharath, Liotsiou, Dimitra, et al. (2018): The IRA, Social Media and Political Polarization in the United States, 2012-2018. Project on Computational Propaganda (https://comprop.oii.ox.ac.uk/wp-content/uploads/sites/93/2018/12/The-IRA-Social-Media-and-Political-Polarization.pdf).

Keller, Franziska B., David Schoch, Sebastian Stier, JungHwan Yang (2020): "Political astroturfing on Twitter: How to coordinate a disinformation campaign." Political Communication 37/2, pp. 256-280.

Kilgarrif, Adam (2015): "Simple Maths for Keywords" (https://www.sketchengine.eu/wp-content/uploads/2015/04/2009-Simple-maths-for-keywords.pdf).

McEnery, Tony, Vaclav Brezina, Helen Baker (2019): "Usage Fluctuation Analysis: A new way of analysing shifts in historical discourse." International Journal of Corpus Linguistics 24/4, pp. 413-444.

Milroy, James, Lesley Milroy (1985): "Linguistic change, social network and speaker innovation." Journal of Linguistics 21/2, pp. 339-384.

Morozov, Evgeny (2011): The Net Delusion: The Dark Side of Internet Freedom. New York: Public Affairs.

"Number of internet and social media users worldwide as of April 2024." Statista, May 22, 2024 (https://www.statista.com/statistics/617136/digital-population-worldwide/).

Paul, Joshua (2021): "'Because for us, as Europeans, it is only normal again when we are great again': metapolitical whiteness and the normalization of white supremacist discourse in the wake of Trump." Ethnic and Racial Studies 44/13, pp. 2328-2349.

Pronoza, Ekaterina, Polina Panicheva, Olessia Koltsova, Paolo Rosso (2021): "Detecting ethnicity-targeted hate speech in Russian social media texts." Information Processing & Management 58/6: 102674 (https://doi.org/10.1016/j.ipm.2021.102674).

Rains, Stephen A., Jake Harwood, Yotam Shmargad, et al. (2023): "Engagement with partisan Russian troll tweets during the 2016 US presidential election: a social identity perspective." Journal of Communication 73/1, pp. 38-48.

Richter, Melvin (1995): The History of Political and Social Concepts: A Critical Introduction. Oxford University Press, USA.

Sadler, Neil (2021): Fragmented Narrative: Telling and Interpreting Stories in the Twitter Age. London:Routledge.

Shu, Kai, Amrita Bhattacharjee, Faisal Alatawi, et al. (2020): "Combating disinformation in a social media age." Wiley Interdisciplinary Reviews: Data Mining and Knowledge Discovery 10/6:e1385 (https://doi.org/10.1002/widm.1385).

"Step-by-step guide to making your first request to the new Twitter API v2", n.d. (https://developer.x.com/en/docs/tutorials/step-by-step-guide-to-making-your-first-request-to-the-twitter-api-v2).

Suk, Jiyoun, Josephine Lukito, Min-Hsin Su, et al. (2022): "Do I sound American? How message attributes of Internet Research Agency (IRA) disinformation relate to Twitter engagement." Computational Communication Research 4/2, pp. 590-628.

"To Russia With Love", April 3, 2020 (https://www.foreignaffairs.com/articles/ukraine/2020-04-03/russia-love).

Traugott, Elizabeth Closs/Richard B. Dasher (2001): Regularity in Semantic Change. of Cambridge Studies in Linguistics. Cambridge: Cambridge University Press.

"Twitter Moderation Research Consortium", n.d. (https://transparency.twitter.com/en/reports/moderation-research.html).

"Wagner chief admits to founding Russian troll farm sanctioned for meddling in US elections", February 14, 2023 (https://edition.cnn.com/2023/02/14/europe/russia-yevgeny-prigozhin-internet-research-agency-intl/index.html).

"Как работали СМИ «фабрики троллей» и что там происходило во время мятежа? Читайте рассказы сотрудников, освободившихся от подписки о неразглашении", July 5, 2023 (https://paperpaper.ru/kak-rabotali-smi-fabriki-trollej-i-ch/).

"Крым", April 26, 2021 (https://www.levada.ru/2021/04/26/krym/).

"Пригожин мертв, но его 'Фабрика троллей' продолжит свою работу", September 1, 2023 (https://newizv.ru/news/2023-09-01/prigozhin-mertv-no-fabrika-trolley-ego-prodolzhit-svoyu-rabotu-418243).

"Путин заявил, что в Российской империи «не было никакой Украины»", November 3, 2023 (https://www.rbc.ru/rbcfreenews/6544ea6d9a7947a51a610793).

"Расследование РБК: Как Из 'Фабрики Троллей' Выросла 'Фабрика Медиа'", March 24, 2017 (https://www.rbc.ru/magazine/2017/04/58d106b09a794710fa8934ac?from=center_1).

Digital Warfare

Strategies and Tactics
A Spinozist Approach to Political Agency Online

Max Kramer

Abstract

Some theorizations of political agency on social networking sites (SNS) has been informed by concepts of "tactics" and "strategies." Such concepts, taken from the language of war, have analytical purchase during a time where "informational warfare" and the infrastructures of digital capitalism have given interactions on SNS a combative and instrumental form. In this article, I will discuss various studies that differentiate between tactics and strategies to comment on how agency on SNS is exercised and conditioned.

By critically engaging with this body of work, I will argue that a theorization of the political is necessary to locate the adequate place for concepts of strategies and tactics. I argue that in many explorations of the concepts of "strategies" and "tactics" there are underlying assumptions about what kind of space the political is that participate in a martial impasse regarding SNS. Furthermore, this impasse is itself driven by the antagonistic qualities of SNS and the lack of institutional regulation. I argue that Baruch de Spinoza's political ethics of ratio and his relational ontology provide ways out of this impasse.

Keywords

Strategies and tactics; online politics; info-wars; Spinoza

Introduction

"This is like a war." I have heard this sentence multiple times during my fieldwork on the digital politics of Muslim activists in India. In response to this "war-like" situation, my interlocutors practice contextual invisibility, retreats and advances, forms of indirect-guarded speech, and anger management. Some strictly separate their private life from their public appearance to avoid their family getting targeted by trolls and street-level, police-backed vigilantes. Others switch platforms once new surveillance regulations and corporate-government collaboration make life on Meta platforms unbearable. They move on whenever the old terrain and the old handle have become either too dangerous or too pointless. All these moves are tactical devices learned from violent online clashes that have, at times, translated into death and rape threats, physical violence and legal harassment. I became increasingly puzzled as to whether the question of how "war-like" social networking sites (SNS) are as a space for minoritized Muslims in India has any bearing on how to conceptualize "tactics" and "strategies" in respect to SNS. And what does this "war-like" condition mean for our understanding of the limits of agency when it comes to politics on SNS?

In recent years, the concepts of "tactics" and "strategies" have received frequent mention in media and communication studies (deRidder 2018; Giraud 2018; Manovitch 2009; Poole et al 2018; see also Schaflechner and Kramer 2024). Although these terms originate from the language of military operations, the aforementioned authors (me included) do not explicitly connect them to ideas of warfare, and those authors who focus on today's information warfare (Singer and Brooking 2019; Jankowicz 2021; Singh and Venkatarayana 2020; Galeotti 2020) do not flesh out concepts of tactics and strategies while using the terms in untechnical ways throughout their work.

The premise of my argument is that the concepts of strategies and tactics carry importance for our understanding of digital politics precisely to the degree that our informational environment is linked to instrumental or weaponized usage: targeted moral outrages, fake news, deep fakes, government-led disinformation campaigns, and so on. The rapid, profit-driven circulation of information and its global reach contribute to a lack of measure both within individuals (e.g. immediate affective reactions) and within political entities (weak regulative institutions), which could enable a less "war-like" scenario on SNS. Hence, the use of the language of tactics and strategies is itself a symptom of an information crisis induced by digital capital (Dean 2010; Stiegler 2019; Fuchs 2021; Carstensen et al 2023) that is compelling digital practitioners and communication scholars alike to conceive of their interactions in military terms – often without reflecting this war-like backdrop.

Thus, tactics and strategies come into play for all those actors bound to, or happy to, operate within an unruly, badly regulated, and outright dangerous terrain where many political interactions are strictly "instrumental", aiming at visibility to accomplish tasks, convey narratives, and to circulate and appropriate affective fragments of meaning (Schaflechner and Kramer 2024). Although there is a flood of info-war manuals, a few social scientists have used the term "information wars" (Singer and Brooking 2019; Jankowicz 2021; Singh and Venkatarayana 2020; Galeotti 2020) to highlight how various tactical and strategic means are utilized by governments, digital mercenaries, troll armies, digital party-volunteers, and non-legacy actors to achieve a range of communicative goals such as destabilizing political and moral orders, expanding geopolitical influence, rebranding the nation-state, commodifying grievances, or creating an environment immersed in a right-wing spectacle surrounding perceived internal enemies.

While the focus of these studies is usually on Russia, the US, or Europe, I'll give examples from India as this happens to be the nation-state I'm most familiar with. But I also have a case for India. Compared with European nation-states India could be considered *avant-garde* when it comes to the depth of neoliberal reregulation, the country's cultural and religious diversity and its ethno-nationalist movement. I claim with the Comaroffs (Comaroff and Comaroff 2015) that India is a place that shows Europe some of its more dangerous political futures. European states have been impacted by decades of neoliberalism, the decomposition of welfare states and media systems, and an increasing range of ethnic, religious, and cultural diversity. The negative emotions set free in this process are exploited in tandem by platform capitalists and far-right populists (Illouz 2023). In India the latter have built a solid hegemony throughout the state and civil society for much longer than what is currently discussed as right-wing populism in Europe (Jaffrelot 1998).

The governing Hindu nationalist Bharatiya Janata Party (BJP, lit. Indian People's Party) regularly orchestrates moral panics in a strategic manner (Singh 2018), streamlining the media system to support the further decomposition of Indian liberal democracy by strengthening executive power in an authoritarian manner (Jaffrelot 2021; Chowdhury and Keane 2021). In the process, Indian Muslims, Christians, indigenous and lower caste activists, and the left opposition have been routinely "othered" throughout the media landscape. All along this streamlining process critical reporters have been targeted (The Wire 2023), some have been killed, leading to India's massive drop in press freedom rating to 150th on RSF's World Press Freedom Index (Reporter ohne Grenzen 2022).

The often embedded and conformist nature of legacy media in India has found its iconic representation in the image of a police-embedded photojournalist stomping in ecstasy on the dead body of a Muslim migrant killed by security forces in the northeastern federal state of Assam in the year 2021 (Singh 2021). The video of the murder of the man, "armed" with a wooden stick, was presumably filmed by a member of the security forces while a camera dangled around the photojournalist's neck as he went about his danse macabre. This became an iconic image of the Hindunationalist state of info-war, with security forces – and not, as is more common, digital vigilantes – filming their brutal exploits, and photojournalists being the first to acknowledge that their job is to participate in the joyous spoils of public violence (Hansen 2021). The situation seems to have reached a stage that can be summed up well by Ford and Hoskins (2023): "[W]ar and its representation have collapsed into each other [...]".

However, the circulation of the image of the photojournalist's dance of death also shows how public violence drives what Ravi Sundaram calls the "Hindunationalist crisis machine" (Sundaram 2020): an atmospheric drone of public violence into which online actors often tap. Most of its spoils are going to the Hindu nationalists who have mastered, like other far-right groups, the art of conjuring realities for propagandistic purposes (Filietz and Marks 2020; Sundaram 2020; Stegemann and Musyal 2020). Sundaram's concept of the "crisis machine", however, goes beyond intentionally produced political propaganda. It points to the way in which a digital infrastructure of smart devices and mobile cameras participate in an atmospheric circulation of public affect (Sundaram 2015). This turns social networking sites in India into spaces saturated with images of atrocity. And it produces an online atmosphere where the strategic and tactical deployments of highly affective imagery of public violence – done to oneself or others – translates into public efficacy: obtaining votes, getting things done (Hansen 2021). This, in turn, sustains an atmosphere that feels much like civil war, especially for the more precarious sections of the Indian population. The Hindunationalist crisis machine keeps this atmosphere at varying temperatures. Most importantly, it can strategically "heat it up" whenever expedient (e.g. before election).

Before I move further into the distinction between strategies and tactics, I first need to address how war-like the political itself is. I do this for the simple reason that the authors I'll discuss concerning tactics and strategies refer to such grounding conceptual debates or have implicit assumptions about how the political is constituted anyways.

War, Peace and the Political – A Spinozist Approach

> For peace is not the mere absence of war, but a virtue based on strength of mind.
>
> (SPINOZA 1958: 311)

Michel Foucault saw "the expression of war" as "invested in the order of political power" (Foucault 1978: 102). According to Foucault, this model allows the struggles that establish institutions to be foregrounded, where "truth functions exclusively as a weapon that is used to win an exclusively partisan victory" (Foucault 2003: 57). For Foucault, an early modern understanding of war-like politics as the institutive dimension has the advantage of providing a "strategic model" (Foucault 1978: 102) that retains the historical nature of knowledge and avoids any universalist certainties to be employed as the foundation of political orders.

Foucault's assessment sharply contrasts with the prevailing view in European political philosophy. This crucially includes rationalist Carl von Clausewitz (2008) who famously considered war as "the continuation of politics by other means". If war continues politics then it does so from a sphere of institutionalized interactions carried out within a state – rationalized through laws, interests, institutions, and (in the liberal add-on) based on regulatory ideas of human rights, moral autonomy and the division of state powers. Foucault calls the entirety of the abovementioned elements by diverse thinkers of various liberal, natural rights and state-rationalist persuasions as pertaining to the "judicial discourse".

However, this distinction between war and law, along with Foucault's own strategic preference for early modern war-like conceptual language, ultimately revolves around two different forms of power: the instituting (*potentia*) and the instituted (*potestas*). Baruch de Spinoza (2020) teaches us that both are crucial for democracy to flourish, and we cannot favor one over the other (Saar 2013; Lordon 2021). Instead of positioning the conceptual language of war and confrontation as a golden possibility to repoliticize, as it was conceived by a number of 20th century European philosophers strongly received in media and communication studies (Schmitt 2015; Laclau and Mouffe 1985; Foucault 1978; Deleuze and Guattari 2013), I will address how information warfare – as one mode of the mediatized condition of political action today – is part and parcel of how social networking sites create an affective infrastructure that fosters war-like "weaponized" exchanges. My premise is that to say SNS "weaponize" information (Singer and Brooking 2019; Galeotti 2020; Ford and Hoskins 2022) means that time, space and affective relations engendered by SNS further interactions that are strictly instrumental. Several studies (Karlsen et al 2017; Udupa 2015; Mukherjee 2022) stress that interactions on SNS often acquire a combative character and for socio-economical, infrastructural and machinic reasons don't resemble public deliberations (Simonowski and Reichert 2020; Fuchs 2021).

In a Spinozist framework, war can be understood as a decomposition of institutional arrangements (within political bodies: civil war; between them: war). If institutions are well-arranged, they can sustain peace. The means to overcome war are to be found in the individual and institutional development of a strength of mind, or *ratio* (see below). With Spinoza we can understand the political as a relation between political bodies, techno-social infrastructures, and individuals' powers to act. Spinoza held that everything exists in relations. He calls the things which stabilize within these relations "modes". Such modes must be understood as power relations, driven by the desire to sustain themselves (*conatus*). The modes (e.g. an individual human being or a political organisation) gain agency by the fortitude of their embodied minds. This fortitude of mind is measured by the degree to which we understand what affects our actions.

For Spinoza, most things aren't all or nothing affairs but rather multi-directional modifications that involve degrees of power and degrees of clarity. It is a cognitivist approach that takes the body seriously: a higher degree of clarity about a state of the body (be it a human body or a political body) amounts to a higher degree of power in respect to it. While it can be perfectly clear to us that triangles add up to 180 degrees, political relations are blurred by the limitations set to our understanding by complexity. There is always too much involved for our minds to grasp, hence taking account of all causalities that factor into the to-be-explained state of affairs isn't humanly possible. This is especially true when it comes to politics.

Therefore, to increase our powers of acting, we will have to figure out the multiple causalities that determine every singular situation to the degree possible by the limited beings we are, affected by all kinds of hostile environmental influences (think of the combination of unemployment, digital capital's interest in profits, and negative emotions circulating on platforms that feed the latter's interest). What Spinoza calls ethics is nothing else but the process to learn about causalities and to find their logical order in our mind. If we do this right, we get a more reliable understanding of our place in life which in turn allows us to give our life a form adequate to our understanding.

Truth, for Spinoza, is not reducible to a strategic model, as it is for many philosophical skeptics and info-war gurus. Instead, it remains unsettled between the universalizing powers of *ratio* and the contextual *imaginatio* (the blurry and limited way of knowing while being affected through external bodies) into which we are thrown by the fact that we aren't very often the adequate cause of our own actions (Norris 1991). This means that I often cannot retrace to a high degree of consistency the steps that ended me up, say, in a conversation with an obnoxious neighbor or in a WhatsApp group that affects me in ways that eat at my mental health.

The concept of *ratio* speaks of forms of emotional subjectivation and has little to do with the division between reason and emotion in the history of European thought. Beth Lord (2018: 61) develops three Latin meanings of *ratio* implicit and interrelated in Spinoza's work: reason, relation, and proportion or measure. Reason is the reconstruction of causalities – as in "finding a reason for why something has happened" (ibid.). For example, the *ratio* of getting imprisoned for my Indian Muslim interlocutors may include having been caught in mediatized moral outrage that has been instrumentalized by their Hindunationalist opponent. Understanding these causalities means understanding things and oneself adequately. Secondly, *ratio* means relation: everything that affects me in the interrelated space I'm living in (e.g. the laws on "seditious speech" in the political body I'm living in may affect my ability to express myself to some degree). Thirdly, *ratio* is a relation of motion and rest that sustains a body (ibid.). This could be called proportion or measure. The proportional movement of our organs sustains the life of an individual human being, while the legitimacy of political institutions – through their power to affect its citizens – sustains the *ratio* necessary to prevent an outbreak of civil war or a virus from decomposing our collective bodies (and, eventually, those of the individuals within it).

Hence, in a Spinozist understanding, it is a matter of context and perspective whether social media can be understood as "war-like" or "a matter of state reason". If, for instance, we consider digital activists, the language of war may resonate strongly with their everyday experiences, and they may employ war-like rhetoric in their digital communications where it seems that every word, including "human rights," gets you to the trenches. One is actively challenging – or trying to institute – new power relations (Spinoza calls this *potentia*) against an enemy.

However, for a regulator in a state institution who upholds instituted power (Spinoza's *potestas*), the language of strategies and tactics is helpful only to the degree that some conflict may be implied and avoidable in their action. The point is to uphold the political body in a de-contested and composed form, and to keep it strong. This also means to uphold relations of motion and rest within the body (its institutional infrastructure) in its currently instituted form, say as a liberal democracy. Some Spinoza scholars (Saar 2013; Lordon 2022; Norris 1991) have convincingly argued that this instituted dimension is not capturing or extracting some ontologically deeper instituting dimension of power as several neo-vitalist commentators hold (Deleuze and Guattari 2013; Negri 2000). These two forms of power, *potentia* and *potestats*, are rather in a necessary inter-relationship.

In a Spinozist approach state bureaucrats may also be seen as operating in tactical and strategic ways, especially when they try to institute new forms of the political from within institutions, against some conservative pushback. However, once they manage to uphold a constituted form, they aren't busy within agonistic or antagonistically structured power relations but rather try to align the affects of the multitude to the *ratio* (the measure of motion and rest) of the state. The regulator's focus is, e.g., on how to regulate affect within a specific population (e.g. hate speech legislation) and within the framework of the state or an international-legal body like the EU.

Tactics and Strategies

The studies dedicated to tactics and strategies on SNS often draw on the theoretical concepts of Michel de Certeau's "Practices of Everyday Life" (1984) and, to a lesser extent, James Scott's "Weapons of the Weak" (1985). Especially de Certeau's distinction between strategies and tactics has been heavily relied upon and is, indeed, useful to think through digital agency. Strategies here refer to practices that shape the space of interaction, while tactics involve creative use of a space that is organized from above. de Certeau (1984) provides the example of a city's layout. If a city administration constructs a functional modernist complex without regard for context, you'll soon see paths that cut across the lawns with people taking shortcuts. Everyday live will remix brutalist paradise.

Although individuals do not have direct influence over city planning, they can navigate creatively within it and leave traces that can guide others' actions. According to de Certeau, a tactic "expects to work on things in order to make them its own [...] or habitable" (ibid. 1984: 5). Another important aspect of a de Certeau's "tactics" is its lack of a centralized structure, which renders its activities a form of subversion that cannot be easily mapped (Manovich 2009). This point is particularly relevant at a time when digital tracking treats anyone as a potential threat in our increasingly securitized societies (Zuboff 2019).

Lev Manovich (2009) is, to my knowledge, the first scholar who brought the discussion of "tactics" to the study of social networking sites and Web 2.0 prosumer culture. In his 2009 essay, he combined questions regarding political economy and infrastructure with those of agency, arguing that

"since the publication of *The Practice of Everyday Life*, companies have developed new strategies that mimic people's tactics of bricolage, re-assembly, and remix. In other words: the logic of tactics has now become the logic of strategies." (Manovich 2009)

According to Manovich, what happens from above resembles increasingly what happened from below (this corresponds to the work of Luc Boltanski and Eve Chiapello [2003] on the artists' critique and the changes in management literature during the transition from Fordist to post-Fordist capitalism). In short, what used to be a tactic becomes a strategy sold as a tactic.

In a study by Sander de Ridder (2015) on teenagers' intimate storytelling about sexuality, the author draws a distinction between tactics and strategies along the lines of de Certeau. While SNS companies may employ strategies to control the space of interaction, non-heteronormative actors use resistant tactics in their everyday lives, which involve subversive pleasures. Building on this, de Ridder adopts Chantal Mouffe's (2013) understanding of the political to explore social networks as spaces of conflict where plural hegemonic struggles take place (originally proposed by Laclau and Mouffe [1985]). In terms of socio-material complexity, de Ridder differentiates between various levels on which the practice of intimate storytelling unfolds, including technology, participation, affordances, representation, and subjectivity (de Ridder 2015: 359). De Ridder emphasizes the political economy of SNSs, highlighting how they prioritize certain information and operate in a top-down manner. Consequently, there exists a "continuous dialectical relation" (de Ridder 2015: 362) between tactics and strategies, which are defined as market powers and cultural powers.

The work by Poole et al. (2021) examines the concept of tactics through an analysis of the far-right hashtag "stopislam". Building on Lezaun's research (2018), they view tactics as "digital weapons of the weak" (Scott's term) that shape the possibilities of social media. Poole et al. specifically explore the limitations of counter-speech when it comes to Islamophobic hashtags. They argue that while certain tactics may subtly alter the possibilities of social media use, these changes often end up amplifying existing inequalities in representation, rather than challenging them (Poole et al., 2021).

Poole et al. caution that individuals who are most affected by these issues can unknowingly exacerbate their situation by circulating counter-speech that is co-opted by their political opponents. Even though Twitter provides space for hate speech to circulate, the platform also creates opportunities for what Bruns and Burgess (2011) call "ad hoc publics" that gather to challenge such sentiments. However, Poole et al (2021) warn that these narratives carry the risk of legitimizing social media platforms' self-presentation as champions of free speech, as they claim to "serve the public conversation" (Twitter 2019).

Eva Giraud explores the concept of tactics within the context of "digital food activism" in what she refers to as "non-hierarchical, leaderless" movements (Giraud 2018, 130). She takes a media-ecological approach that goes beyond the specific platforms used (Madianou and Miller 2013, 170) and emphasizes the significance of "frictions between technologies, activists, and issues" (Giraud 2018, 130). Giraud argues that these frictions have the potential to complicate theoretical narratives about the commodification of dissent (Giraud 2018, 131). Giraud takes us on a longer journey, exploring various conceptual approaches to understanding socio-technical assemblages from a media ecological perspective. Eventually, she considers the particular value of understanding activism from a media ecology perspective as a way to explore how the affordances of specific media technologies shape "non-hierarchical modes of politics" (Giraud 2018: 141). The examples she provides regarding tactics mostly involve "tinkering." Giraud's point is that actions like "hashtag politics," dismissed by some theorists (Dean 2009; Stiegler 2021) as forms of self-promotion and liberal individualism characteristic of communicative capitalism, have some limited scope when understood as part of a communication ecology (Giraud 2018: 143). This occurs when social media is connected to more radical media practices and enviroments, where ephemerality and aggregative logics transform into a more sustained political identity (Giraud 2018: 143).

An Arrangement of Time-Space and Affect

In India, the lack of a political body that could sustain the political expression of minorities merges with some war-like qualities inherent in digital networks. It is precisely from this perspective that we should be critical of claims regarding the ontologically antagonistic nature of the political – the antagonism is rather the result of the lack of a peaceful political body under the conditions of global financial capital.

Let me give an example from India to delineate the kind of problematic racialized Muslim practitioners find themselves in when using SNS. Ahmed, an Indian Muslim activist, has been caught in several moral panics marking him as a violent Muslim man hellbent on destroying the integrity of the Indian nation. These staged moral panics made it difficult for him to sustain his digital practice. He was imprisoned, saw his public persona appropriated by political opponents, regularly received death threats, and one attempt on his life was made. Moving tactically, in his case, implies working on his emotions in order not to get caught in the digital infrastructures that mold time into immediate reactions and hold a set number of stereotypical emotions in store for racially marked subjects like him. There are "emotional orders" (Stodulka 2019) that to some degree define which subjects are in need of regulation as they are perceived as potentially violent. The Indian Muslim male has been thematized as one such emotionally out-of-control subject – once he is visible as a Muslim. In the case of Ahmed, an anti-terror squad picked him up from home as he was planning a non-violent street protest against a discriminatory citizenship law.

Once released from jail, he posted a line of an Urdu poem that contained the word "ashes". Following the Hindunationalists' logic of emotional and agentive ascription, ashes follows fire, fire follows riots, and is linked to anger. Angry Muslim men are seen as violent and a threat to the nation. A poet's line was dealt with penal law. The problem of tactics for Ahmed is both affective (working on his emotional dispositions in respect to the societal ordering of emotions, relating to the affective relationality of Twitter) and spatio-temporal. An immediate response makes you part of events – or rather: part of you becomes part of an event –, it modifies the bio-social rhythms that sustain you and your public persona by reassembling parts and pieces of subjectivity to fit the event along the logics of racialized digital capital. What fits moral panics are spatial elements: the reproduction of tropes or commonplaces. But Ahmed is also confined by his physical location in India in ways that Indian Muslim digital practitioners in Dubai are not (they don't have to fear the police barging into their homes for a tweet that quotes a poem, at least as long as it is related to India).

Ahmed found ways to use the affordances of Twitter so that it became difficult to get captured in moral panics. He did so by focusing on the documentation of Anti-Muslim violence, without personal statements attached, by creating a strictly public persona (no private information, always wearing the same 'public dress' etc.), by not using addressivity markers such as @ and # that lead to fast-paced pitched battles. He instead builds the strength of his tactical intervention around his reach (+100k followers), moral persona, and the qualities of audio-visual material to undermine the Hindu nationalist frame. All the while, it is clear to him that he is being kept ready for electoral mobilization via moral panics. If the timing of his posts isn't right (a poem containing "ashes" after release from prison), if some tropes are touched upon (e.g. ironically), he may get into serious trouble.Pace Lev Manovich's important essay from 2009, the tactics used on social networking sites (SNS) are no longer primarily focused on modifying symbols. This is because the realm of public visibility is dominated by spectacle (Briziarelli and Armano 2017), surveillance (Zuboff 2019), and moral outrage (Crockett 2017). Users like Ahmed move online within regimes that combine elements of disciplinary power (Foucault 2019) and the power of control (Deleuze 1992). The former seeks to normalize and stabilize our identities, confining subjects within institutions and positions, while the latter attempts to capture us on the move. Control-societies measure future behavior, dispersing our subjectivities into what is trackable and what is not (Mühlhoff et. al 2019), into what can be aligned to more powerful desires and what not (Lordon 2014). I argue that the concepts of strategy and tactic need to be further abstracted: strategy means structuring space-time and affect to align subjects, or the willing parts of bodies, to one's desire. Tactics, on the other hand, are short-term, reactive arrangements of time-space and affect. They involve no control over space but a reflexive movement aware of its traceability, its affective relationships, and its possible effects resulting from the constantly changing affordances that SNS offer in the institutional arrangement a practitioner finds herself within (especially the nation-state's regulation and policing of digital space). In this sense, strategies allow for *potentia*, for an active restructuring of power relations, while tactics allow for moves within *potestas* (already established power-relations). The latter operate with degrees of passivity that can easily lend themselves to the expression of negative emotions and circles of sacrificial self-destruction. Negative emotions are one of the driving forces of far-right crisis machines worldwide as they converge with the data hunger of digital capital and may trap minoritized subjects in the very orders they are struggling against. This is the problem of capture that I'll sketch in the next subsection.

Being Tactical Means Being Aware of Capture

The concept of capture speaks of socio-machinic and cybernetic processes that form and circulate "information" and valorize affect and symbols in historically sedimented and often racialized ways (Udupa and Dattatreyan 2023; Schaflechner and Kramer 2024).

The concept of capture helps us understand the limits of digital agency in response to ongoing economic and techno-social transformations where "[...] everything recorded of you or about you as you go about your daily life is captured by a vector and fed into computation to figure out how better to use you for the greater glory of Amazon, Google, Apple, or some other company" (Wark 2019, 11). Following McKenzie Wark, I think that today's economic and technological mediations of subjectivity call for a vectorial analysis. In this sense, practices can be seen as responsive to an affective ecology (Angerer 2017; Brejak and Mühlhoff 2019) that aligns subjects by working on their desires and affections (Lordon 2014; Citton 2016). This makes it necessary to approach tactics as trajectories under the threat of capture. Capture has two functions. In the first, we're becoming aligned with the desires useful to digital capitalists (e.g., moral outrage), and second, we fall into what Udupa and Dattatreyan (2023) call a "recursive trap" that allows for all kinds of political appropriations (cynical and strategic from the far-right; see Stegemann and Musyal 2020) and security measures. Such securitizing regimes have been shown to continue historically sedimented power relations that are reinscribed in algorithmic racism and through digital control mechanisms (Ahmed 2007; Udupa and Dattatreyan 2023).

This, precisely, is the next good reason to read Spinoza again these days. After all, his way of approaching agency initiated the vectorial approach: "I shall consider human actions and appetites as if the subject were lines, surfaces, or solids" (Spinoza 2020, 162). There is no machinism here; what we are dealing with is rather a way of reconstructing affect and causality through geometry. Although it was ontological for Spinoza, we can take it, instead, as a heuristic for the kind of media anthropological analysis that takes political bodies, political economy, and individuals' political-ethical practices seriously as power relations (Saar 2013) while abstracting enough from everyday interactions to not invest in some ready-made, traditional, or moralizing notion of the "good" subject or the "good" practice.

Conclusion

Although Spinoza did not operationalize a concept of tactics himself, he moved tactically through much of his life following the motto *caute* (caution). His caution was a blend of thinking and acting to the highest degree of self-causation in the limiting environment given to him. All the while he did not let go of what rationally called him to action: his understanding of an active God. After all, Spinoza was a political radical intensely disliked by the orthodoxies of his day, banned from his own Jewish Sephardi community and outrightly demonized in the century following his death (Israel 2002; Nadler 2018).

In this article, I make a couple of additions to de Certeau's influential notion of tactics by drawing on my work on Indian Muslim digital practitioners and a Spinozean line of theorization. In one fundamental aspect I agree with de Certeau and the research on digital tactics that followed him: tactics are short-term adaptations and appropriations in a space that is not your own and that you can't structure yourself. However, I want to give more contours to the concept of tactics in digital "info-war"-settings such as those Indian Muslims find themselves within. I argue that tactics need to be understood as material "arrangements" of time-space and affect in constant threat of capture. Capture is the process that sucks in and cybernetically transforms affect, spatiality, and temporality in ways that serve capital. It reinforces racialized emotional-moral orders and security regimes aligned to them.

Furthermore, I follow Spinoza in seeing tactics as reactive modes of action – there is never a sufficient degree of self-causation possible on commercial social networking sites (they aren't expressions of the multitude). A tactically operating subject knows how to stay below the threshold of capture. It operates with the highest degree of self-causation that is possible within a reactive space, for example by being capable of utilizing SNS' reach to bring forth marginalized perspectives.

Lastly, I argue that the political should not be confused with universal ideas such as "bottom-up resistance", "antagonism", "rational principles" or "spaces of appearance and recognition". Such ideas, if not theorized through affective relations, may distort the coordinates of tactics and strategies that empirical research needs to flesh out in the first place. This is especially true regarding regions such as South Asia and groups like Indian Muslims that haven't received much attention so far and may be relevant to rethinking our explanatory models.

Funding

This research is funded by the Volkswagen Foundation and conducted within the project "The Populism of the Precarious", Freie Universität Berlin.

References

Ahmed, Sara. 2007. 'A Phenomenology of Whiteness'. *Feminist Theory* 8 (2): 149–68. https://doi.org/10.1177/1464700107078139.

Angerer, Marie-Luise. 2017. *Ecology of Affect: Intensive Milieus and Contingent Encounters*. Lüneburg: meson press eG.

Boltanski, Luc, and Eve Chiapello. 2003. *Der neue Geist des Kapitalismus*. Translated by Michael Tillmann. Konstanz: UVK.

Briziarelli, Marco, and Emiliana Armano, eds. 2017. *The Spectacle 2.0: Reading Debord in the Context of Digital Capitalism*. University of Westminster Press. https://doi.org/10.16997/book11.

Bruns, Axel, and Jean Burgess. 2011. "The Use of Twitter Hashtags in the Formation of Ad Hoc Publics," August.

Carstensen, Tanja, Simon Schaupp, and Sebastian Sevinani. 2023. *Theorien Des Digitalen Kapitalismus*. Frankfurt am Main: Suhrkamp.

Certeau, Michel de. 1984. *The Practice of Everyday Life*. 1. print. Berkeley, Cal. [u.a.]: Univ. of California Pr.

Chowdhury, Debasish Roy, and John Keane. 2021. *To Kill A Democracy: India's Passage to Despotism*. Illustrated edition. Oxford, United Kingdom: Oxford University Press.

Citton, Yves. 2016. *The Ecology of Attention*. 1st ed. Cambridge Malden, MA: Polity.

Clausewitz, Carl von. 2008. *On War*. London: Penguin.

Comaroff, Jean, and John L. Comaroff. 2015. *Theory from the South: Or, How Euro-America Is Evolving Toward Africa*. London: Routledge.

Crockett, M. J. 2017. 'Moral Outrage in the Digital Age'. *Nature Human Behaviour* 1 (11): 769–71. https://doi.org/10.1038/s41562-017-0213-3.

De Ridder, Sander. 2015. "Are Digital Media Institutions Shaping Youth's Intimate Stories? Strategies and Tactics in the Social Networking Site Netlog." *New Media & Society* 17 (3): 356–74. https://doi.org/10.1177/1461444813504273.

Dean, Jodi. 2010. *Blog Theory: Feedback and Capture in the Circuits of Drive*. 1st edition. Cambridge, UK ; Malden, MA: Polity.

Deleuze, Gilles. 1992. "Postscript on the Societies of Control." *October* 59: 3–7.

Deleuze, Gilles, and Felix Guattari. 2013. *A Thousand Plateaus: Capitalism and Schizophrenia*. 1st ed. London New York Oxford New Delhi Sydney: Bloomsbury Academic.

Dolata, Ulrich, and Jan-Felix Schrape. 2023. "Politische Ökonomie Und Regulierung Digitaler Plattformen." In *Theorien Des Digitalen Kapitalismus*, 344–63. Frankfurt am Main: Suhrkamp.

Ford, Matthew. 2023. *Radical War*. London: Hurst.

Foucault, Michel. 1978. *The History of Sexuality*. New York: Pantheon Book.

———. 2003. *Society Must Be Defended: Lectures at the Collège de France 1975-1976*. New York: Palgrave Macmillan.

—. 2019. *Ethics: Subjectivity and Truth: Essential Works of Michel Foucault 1954-1984*. New Ed Edition. London: Penguin.

Fuchs, Christian. 2021. *Das digitale Kapital: Zur Kritik der politischen Ökonomie des 21. Jahrhunderts*. 1st edition. Wien Berlin: Mandelbaum Verlag eG.

Galeotti, Mark. 2023. *The Weaponisation of Everything: A Field Guide to the New Way of War*. New Haven: Yale University Press.

Giraud, Eva Haifa. 2018. "Displacement, 'Failure' and Friction: Tactical Interventions in the Communication Ecologies of Anti-Capitalist Food Activism." *Digital Food Activism*, 130-.

Hansen, Mark. 2012. "Ubiquitous Sensation or the Autonomy of the Peripheral: Towards an Atmospheric, Impersonal and Microtemporal Medi." In *Throughout: Art and Culture Emerging with Ubiquitous Computing*, edited by Ulrik Ekman, 63–88. Cambridge: MIT Press.

Hansen, Thomas Blom. 2021. *The Law of Force: The Violent Heart of Indian Politics*. New Delhi: Aleph Book Company.

Israel, Jonathan I. 2002. *Radical Enlightenment: Philosophy and the Making of Modernity 1650-1750*. Illustrated Edition. Oxford New York: Oxford University Press.

Jaffrelot, Christophe. 1998. *The Hindu Nationalist Movement In India*. 0 Edition. New York: Columbia University Press.

—. 2021. *Modi's India – Hindu Nationalism and the Rise of Ethnic Democracy*. Chennai: Context.

Jankowicz, Nina. 2021. *How to Lose the Information War: Russia, Fake News, and the Future of Conflict*. London New York Oxford New Delhi Sydney: I.B. Tauris.

Karlsen, Rune, Kari Steen-Johnsen, Dag Wollebæk, and Bernard Enjolras. 2017. "Echo Chamber and Trench Warfare Dynamics in Online Debates." *European Journal of Communication (London)* 32 (3): 257–73. https://doi.org/10.1177/0267323117695734.

Kramer, Max. 2022. "Beyond the Identitarian Deadlock: Why Mobile Methods Are Useful for Studying Media in Zones of Conflict." *Studies in Indian Politics* 10 (2): 289–97.

Lord, Beth, ed. 2018. *Spinoza's Philosophy of Ratio*. Edinburgh: Edinburgh University Press.

Lordon, Frédéric. 2022. *Imperium*. London: Verso.

—. 2014. *Willing Slaves Of Capital: Spinoza And Marx On Desire*. London: Verso Books.

Manovich, Lev. 2009. "The Practice of Everyday (Media) Life: From Mass Consumption to Mass Cultural Production?" *Critical Inquiry* 35 (2): 319–31.

Mouffe, Chantal. 2013. *Agonistics: Thinking The World Politically*. London ; New York: Verso.

Mouffe, Chantal, and Ernesto Laclau. 1985. *Hegemony and Socialist Strategy: Towards a Radical Democratic Politics*. London: Verso Books.

Mühlhoff, Rainer, Jan Slaby, and Anja Breljak. 2019. *Affekt Macht Netz: Auf dem Weg zu einer Sozialtheorie der Digitalen Gesellschaft*. Bielefeld.
Mukherjee, Rahul. 2020. "Mobile Witnessing on WhatsApp: Vigilante Virality and the Anatomy of Mob Lynching." *South Asian Popular Culture* 18 (1): 79–101. https://doi.org/10.1080/14746689.2020.1736810.
Nadler, Steven. 2018. *Spinoza: A Life*. 2nd ed. Cambridge ; New York: Cambridge University Press.
Negri, Antonio. 2000. *Savage Anamoly: The Power of Spinoza's Metaphysics and Politics*. Minneapolis Oxford: University of Minnesota Press.
Poole, Elizabeth, Eva Haifa Giraud, and Ed de Quincey. 2021. "Tactical Interventions in Online Hate Speech: The Case of #stopIslam." *New Media & Society* 23 (6): 1415–42.
Ramdev, Rina, Sandhya Devesan Nambiar, and Debaditya Bhattacharya. 2015. *Sentiment, Politics, Censorship: The State of Hurt*. New Delhi: SAGE Publications India Pvt, Ltd.
Saar, Martin. 2013. *Die Immanenz der Macht: Politische Theorie nach Spinoza*. Originalausgabe Edition. Berlin: Suhrkamp Verlag.
Schaflechner, Jürgen, and Max Kramer. 2024. "Tactics for Becoming Visible." *Dialectical Anthropology* 48 (1).
Schmitt, Carl. 2015. *Der Begriff des Politischen.: Text von 1932 mit einem Vorwort und drei Corollarien*. 9., Korrigierte Edition. Berlin: Duncker & Humblot.
Schroeder, Ralph. 2018. *Social Theory After the Internet: Media, Technology and Globalization*. 1st ed. London: UCL Press.
Scott, James C. 1985. *Weapons of the Weak: Everyday Forms of Peasant Resistance*. New Haven: Yale University Press. http://archive.org/details/weaponsofweak-eveoooscot.
Singer, P. W., and Emerson T. Brooking. 2019. *Likewar: The Weaponization of Social Media*. Reprint Edition. Boston (Mass.): Mariner Books.
Singh, Namita. 2021. "Photographer Arrested for Stomping on Body of the Protester in India." *The Independent*, September. https://www.independent.co.uk/asia/india/assam-photographer-stomps-dead-protester-b1926310.html.
Singh, Shivam Shankar. 2019. *How to Win an Indian Election: What Political Parties Don't Want You to Know*. Ebury Press.
Simanowski, Roberto, and Ramón Reichert. 2020. *Sozialmaschine Facebook*. Berlin: Matthes & Seitz.
Slaby, Jan, and Christian von Scheve. 2019. *Affective Societies: Key Concepts*. 1st edition. Routledge.
Spinoza, Baruch de. 1958. *Political Works*. Oxford: Clarendon Press.
———. 2020. *Ethics*. Princeton: Princeton University Press.
Stegemann, Patrick, and Sören Musyal. 2020. *Die Rechte Mobilmachung*. Berlin: Econ.
Stiegler, Bernard. 2019. *The Age of Disruption: Technology and Madness in Computational Capitalism*. 1st ed. Cambridge, UK Medford, MA: Polity.

Stodulka, Thomas. 2019. 'Orders of Feeling'. In *Affective Societies*. London: Routledge.
Sundaram, Ravi. 2015. "Post-Postcolonial Sensory Infrastructure." *E-Flux*, 2015. https://www.e-flux.com/journal/64/60858/post-postcolonial-sensory-infrastructure/.
——. 2020. "Hindu Nationalism's Crisis Machine." *HAU: Journal of Ethnographic Theory* 10 (3): 734–41. https://doi.org/10.1086/712222.
The Wire. 2023. "At Least 194 Journalists Were Targeted Across India in 2022: Report." The Wire. June 27, 2023. https://thewire.in/rights/at-least-194-journalists-were-targeted-across-india-in-2022-report.
Udupa, Sahana. 2016. "Archiving as History-Making: Religious Politics of Social Media in India." *Communication, Culture & Critique* 9 (2): 212–30. https://doi.org/10.1111/cccr.12114.
Udupa, Sahana, and Ethiraj Gabriel Dattatreyan. 2023. *Digital Unsettling*. New York: New York University Press.
Wahl-Jorgensen, Karin. 2018. *Emotions, Media and Politics*. 1st ed. Cambridge, UK; Medford, MA, USA: Polity.
Wark, McKenzie. 2019. *Capital Is Dead: Is This Something Worse?* London; New York: Verso.
Zuboff, Shoshana. 2019. *The Age of Surveillance Capitalism: The Fight for a Human Future at the New Frontier of Power: Barack Obama's Books of 2019*. Main edition. London: Profile Books.

When the war goes viral
Ukrainian and Russian War Memes[1]

Elena Korowin

Abstract

In Ukraine's information war against Russia, memes became one of the most powerful weapons in 2022. Dirk von Gehlen describes memes as "catchy tunes and attention machines on the Internet" that are based on the banal principle of copying, adapting, and referencing. (von Gehlen 2020: 16)

Another characteristic of memes is self-empowerment through humor – they are usually easy to receive and close to the audience, memes are vital and democratic. (ibid:7) This makes them interesting for political purposes and an increased emergence of political memes has been observed in the recent past. They work like signs held up at demonstrations, only their reach and distribution can be much more effective. (cf. Sommavilla 2022)

Keywords

Memes; memefication; propaganda; misinformation; parody; trolling

Putin memes

We remember: In the early 2010s, Vladimir Putin was a meme star. He flew with cranes, showed himself to be an experienced judo fighter, rode horses barechested, waded through tall grass with a rifle in hand, dived for antique amphoras and cuddled cute puppies.

The Russian president's trashy-looking official self-portrayal in the media was a perfect template for meme-spreading and was enthusiastically replicated on the Internet. (Schomowa 2020) While in the West the Putin memes were used for sarcastic amusement and at times achieved a real cult status, in Russia, despite their exaggeration, they were also used to express enthusiasm for Putin as a "doer" and as an action hero. The idea of a strong politician, of a "real" man in power,

---

[1] This text is a revised and supplemented version of a published passage from this source: Elena Korowin: *Krieg geht viral. Visuelle Kultur und Kunst im Ukraine-Krieg*, transcript Bielefeld 2023, 109–139.

after the political disasters with Boris Yelzin, was successfully served by Putin's PR strategies. Merchandise sold throughout Russia, catering to both camps: those who wanted to make tongue-in-cheek fun of Putin by wearing a T-shirt with his likeness and those who truly revered him as Russia's savior. (cf. Bidder 2016) Since the protests following Putin's re-inauguration in 2012, memes have changed, becoming more critical and losing their lightness and naivete. However, Putin retained the "Chuck Norris image" of the invincible, all-knowing, and all-powerful president. His internet presence seemed gigantic; between 2013 and 2016, *Forbes Magazine* named him the most powerful man in the world three times in a row. (Vollmer 2014) Most in Russia and outside laughed or were indifferent. What was overlooked was that the ironic exaggeration has increased Putin's power and cult and not everyone can understand irony.

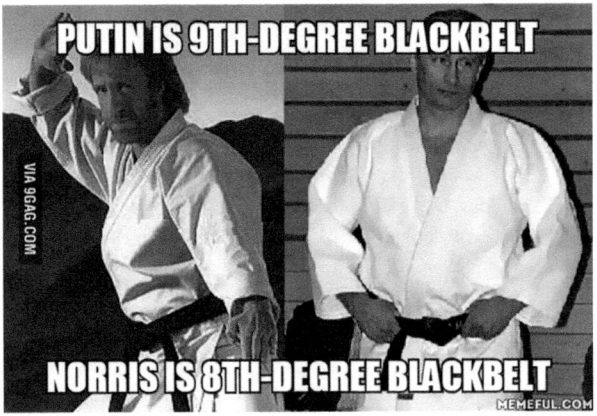

Figure 1: *Example of Putin memes from the early 2010s*

After the annexation of Crimea in 2014, the jokes in Russian propaganda continued, some of them developed into legends about Ukraine and the people of Donbas. Almost no one in Russia seemed, at least publicly, to care about the unscrupulous distortion of the facts. What's more, a large part of the population believed what they were told about Ukraine. (cf. Medvedev 2022) This development was viewed critically on both sides. The Ukrainian author Georgii G. Potschepzov coined the term "War of Perception" in early 2000s, which is applicable to the battle of interpretation over Ukrainian history as well as identity. (cf. Potschepzov 2000) Basically it is comparable to the concept of information warfare. In his sharp analyses, the Russian political scientist Sergei Medvedev shows, with a touch of irony, how the propaganda on Russian television itself affects Putin. The president and the people are beginning to believe what was previously produced as lies by the editors. (cf. Medvedev 2017; Zygar 2023)

In 2017, the exhibition *Putin – a Meme* was shown in Moscow on the occasion of Putin's 65th birthday. The exhibition was organized by three Telegram channels. It showed the 50 most popular memes from 2008 to 2017. (Masalzeva 2017) That same year, Trevor Noah opened a satirical library of Trump's selected tweets in New York. Not to be forgotten is the statement made by a 4Chan participant after Trump's election victory in November 2016: "We made a meme the president". (von Gehlen 2020: 52) Shomova only sees a major turning point in the Putin memes in 2018, when Putin's rhetoric became more aggressive, his re-election as president, the lack of alternatives and his irremovability came to the fore. Interestingly, her report makes no mention of the annexation of Crimea and the events in Donbass since 2014. Neither on the ban on memes depicting famous Russian personalities, which was imposed by the regulator Roskomnadzor in April 2015. ("Russland verbietet Memes. Spaßbremse des Internets" 2015) The Russian government thus officially acknowledged the impact of meme culture with this ban.

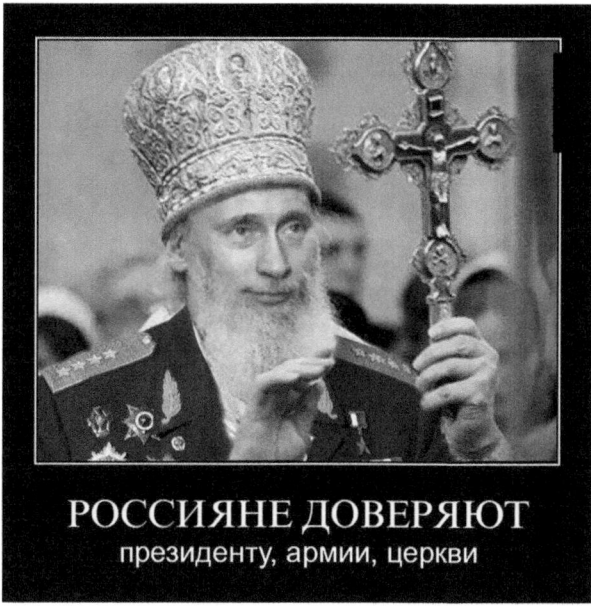

Figure 2: THE RUSSIANS TRUST: the president, the army, the church

Zelenskyy memes

Putin as a meme lost the sympathy of the Internet community by 2018 at the latest and some observers outside of Russia were able to see how all Putin topoi were transferred from satire to reality. (cf. Medvedev 2017) Since the COVID-19 pandemic, the Russian president gradually disappeared from the public eye, rumors spread about his health, and finally he was deposed from his meme throne within a few days in 2022. Volodymyr Zelenskyy, the former controversial president of Ukraine, became the new superhero of meme culture. Since Russia's attack on Ukraine in February 2022, Zelensky has adopted many of the characteristics previously used by Putin but transferred them from the outdated imperial rhetoric of power into a democratic, "woke" present. A few examples: Zelenskyy also shows himself to be a doer, a "real man" – but he doesn't have to artificially create a special setting – keyword: Putin and amphoras that were sunk in the water especially for him.

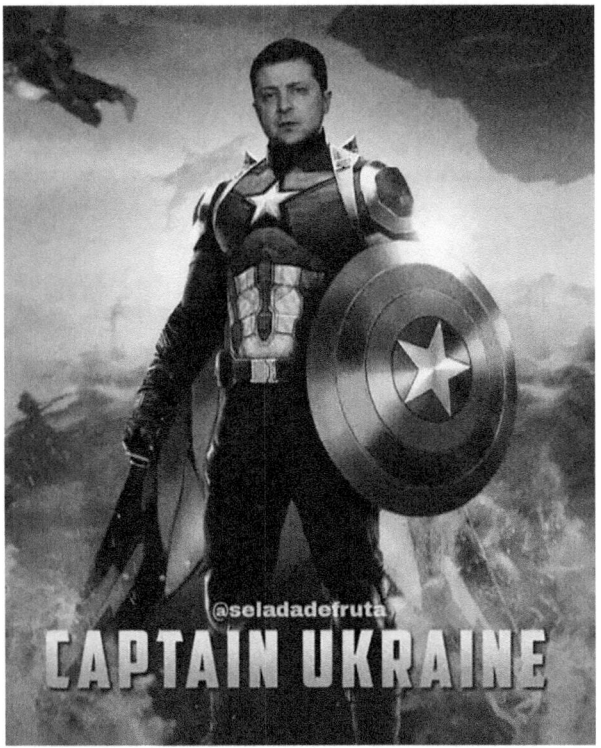

Figure 3: Volodymyr Zelenskyy as Captain Ukraine

The Russian president has created the best conditions for Zelenskyy's memeization. The Ukrainian president reacts to the events taking place around him. When Russia spread false information about his escape from the capital, he promptly showed up in front of the government palace in Kyiv. His entire display of power is based on visibility, which can be crucial in times of war. His Twitter account allowed followers to track where he was and who he was meeting, and this was an important stabilizing factor for the Ukrainian population in the first months of the war. The cultural scientist Annekathrin Kohout describes images on social media as "affordant", they demand a reaction, which means that users must react to them by liking, sharing or clicking them away (cf. Kohout 2022) – Zelenskyy himself also became "affordant".

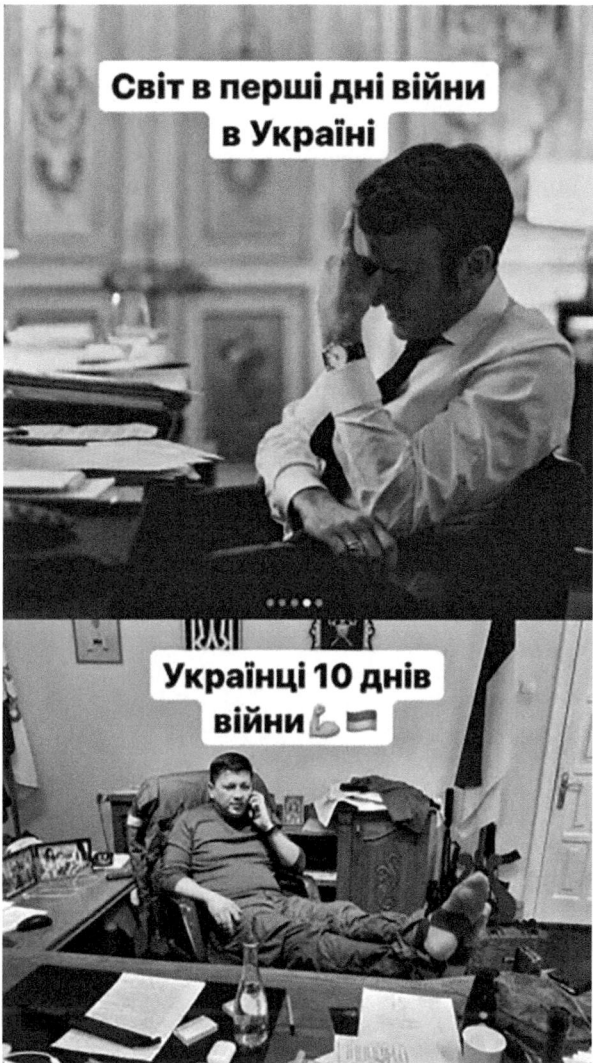

Figure 4: "The world in the first days of the war"/"Ukrainians 10 days of war"/"That's him"

He also perfectly serves the image of a humble hero – in his military t-shirt that has become cult, he appears ready for battle, agile, youthful, strong and forms a contrast to the distant, old and shapeless Putin with his face puffed up by Botox and fillers. (cf. Shuster 2024) This contrast is a perfect starting point for memefication – there are countless comparison memes of both leaders, which are often fed into existing narratives: Zelenskyy as a superhero and Putin as a villain, in any superhero saga. Zelenskyy's success is based on people's need for real heroes. The professional entertainer knows how to deal with this – he has already written himself into the positive hero figure with his television series *Sluha narodu* (Servant of the People 2015–2016). The Ukrainian president is a media professional and knows the importance of the audience and the attention of the masses. He becomes an influencer, so that the otherwise dapper Emmanuel Macron suddenly shows up in the Elysée Palace in a hoodie. (Ullrich 2022) This is where the political future appears, while Putin, who clearly represents the past, becomes invisible. After many years of self-representation, Putin seems to have found his way back to his roots as a KGB man from the particularly strict Leningrad school. With Zelenskyy and Putin, the scissors of populism are establishing a dichotomy: man of the people against leader of the people. The memes depict this in all its facets: Putin at his long neoclassical table in Empire style seems to be reenacting the fate of Napoleon and Zelenskyy, who, as a fighter in casual clothing, takes over and perfects Putin's old self-dramatization. Zelenskyy is clearly the new Chuck Norris of politicians; he has taken over this baton directly from Putin.

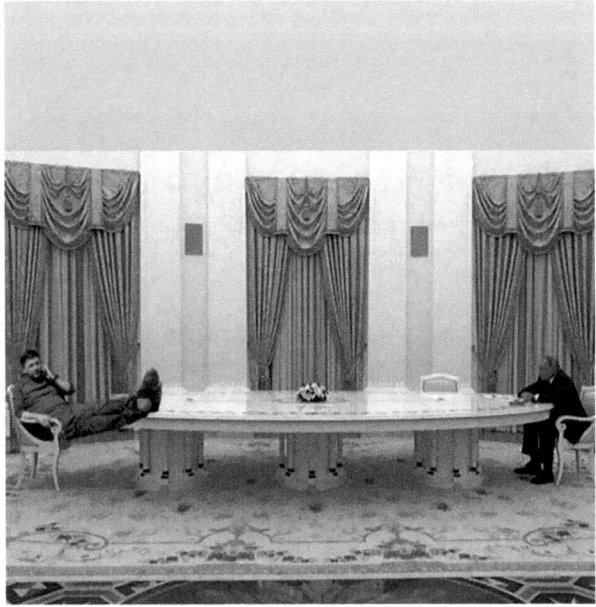

Figure 5: Vitalii Kim visits Moscow

Zelenskyy is making media history with his media presence and staging; his messages use classic professional cinema aesthetics; they are carefully thought out. It puts pressure on you – he is as "affordant" as a social media image. After his posts and video messages, we must act. However, it is not always appropriate to respond based on affects triggered by images. Even image professionals cannot assess this right now. In the Syrian war, emotional witness was created through blurred images; people were convinced that these images were more authentic. Such pictures are rare or non-existent in the war in Ukraine – most of the images are very professional and the quality of the cell phone photos is not comparable to previous ones.

On Zelenskyy's account, the images are reminiscent of film trailers – there has been enormous professionalization here and this can trigger a cognitive dissonance that makes viewers from Europe vulnerable. In Ukraine, in the first months of the 2022 war, there were other superheroes who competed with Zelenskyy in the media, such as Vitalii Kim, the governor of Mykolaiv Oblast.

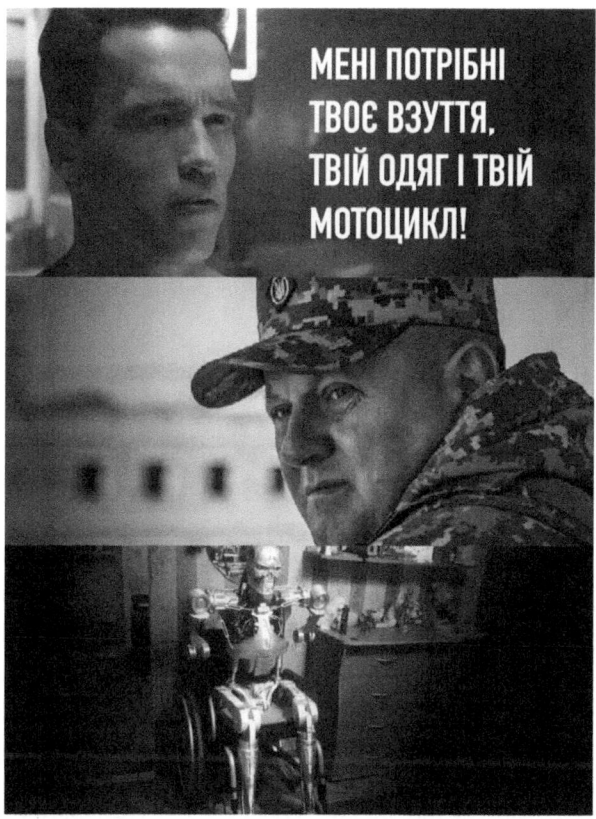

Figure 6: "I need your shoes, your clothes and your motorcycle"

Kim has been sharing daily video updates and spots about the Russian army during the attacks on his territory. And Kim's image has eventually become a popular meme. It was based on a photo that showed Kim wearing casual clothes, feet in colorful socks on the table, talking on the phone in his office. Kim became the epitome of Ukrainian calm in the face of Russian threats. Another hero is certainly the Defense Minister of Ukraine, Oleksii Reznikov, but above all the Commander-in-Chief of the Ukrainian Armed Forces, Valerii Zalushnyi, who, even more than Zelenskyy, serves the image of a warrior and action hero. All these heroes and memes are extremely important for the Ukrainian population as motivation to persevere. They have the same function as agitprop of the 20th century, such as posters and flyers, and in Eastern European countries also Lubki and Rosta windows. A comprehensive analysis of this rich meme material, its meaning and background remains to follow.

Russian warship f*** you!

Today we know how radically February 24, 2022 changed Europe. While people in Ukraine have been confronted with the Russian government's unscrupulousness since 2014 and continued to illustrate it with the gallows humor of memes, worldwide memeization seemed inappropriate in view of the invasion of Russian troops. The question arose: who is allowed to make memes? (Eilenberger 2022; cf. Semotiuk 2023.)

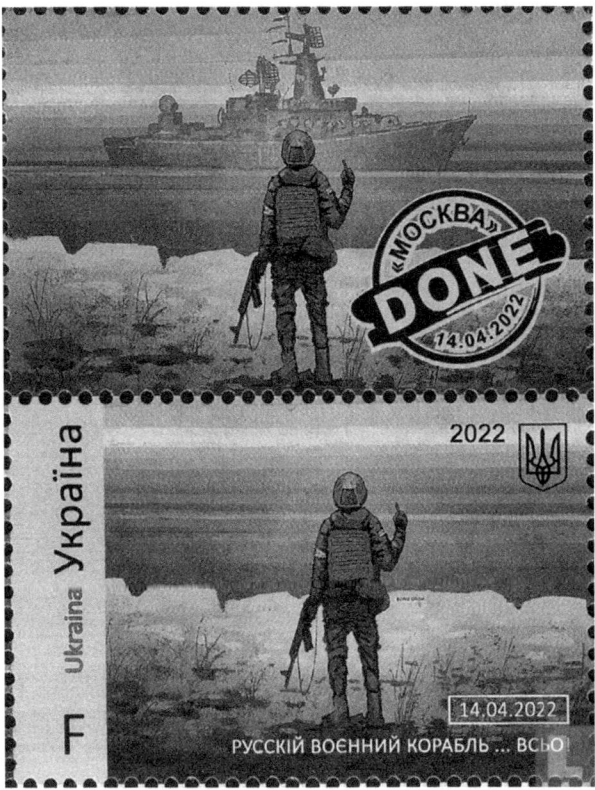

Figure 7: First draft of Boris Grokh's stamp with the warship "Moskva" and second draft with the sunken ship under the title "Done".

The first day of the war brought Ukraine its first effective meme of self-empowerment and a weapon in information warfare. On February 24, 2022, a Russian warship approached Zmiinyi Ostriv (Snake Island) in the Black Sea, which has an area of 17 hectares and a population of 30 people. The 13 border guards stationed on the island were asked to surrender twice: "I am a Russian warship. Suggest laying down arms and surrendering to avoid bloodshed and unjustified casualties. Otherwise, you will be bombed". This request, which is quoted differently depending on the reporter, led to the clear answer from soldier Roman Hrybov, which went viral within hours. (cf.: Balforth/Trevelyan/Jones 2022; Visontay 2022; Sauer 2022)

It became a slogan against Russian aggression – on mugs, shirts, cell phone cases, murals, graffiti, street signs and billboards. The community-building effect of this slogan led to improvised flash mobs via internet memes and appeared on posters around the world. The Ukrainian railway got involved and demonstratively broke up many connections to Russia with the words "Russian train fuck you" (Petrenko 2022).

Hrybov's saying became a unifying slogan against Russian Z-propaganda, which spread similarly quickly on the Internet, on the streets of Russia and within the Russian-speaking diaspora worldwide. On March 12, 2022, Ukrpochta, the Ukrainian Post, commissioned the stamp designed by Borys Grokh, which has now become a collector's item. (Mishchenko 2022) On the anniversary of the Russian attack in 2023, Ukrpochta issued another stamp – this time with a motif by street artist Banksy.

In November 2022, he left his works in several destroyed cities in Ukraine. In Borodianka, one work showed a boy throwing an adult judo fighter to the ground, a clear reference to the Russian president. The motif was printed as a stamp with the signature "ПТН ПНХ" (FCK PTN: short for "Putin, fuck you"). In addition to emphasizing the narrative of David versus Goliath and good versus evil, the Russian warship meme provided a metaphor for the situation of the Russian population. The months since the 2022 invasion have shown that the Russians (voluntarily and involuntarily) are truly sitting in a huge warship led by a dictatorial captain. Many have now jumped overboard.

The Cossacks

In Ukraine, Russia and many other former Soviet states that had a centralized common "cultural history" imposed on them, there is a visual counterpart to the biblical David-Goliath narrative. It is a picture by the Ukrainian painter Ilya Repin, who was one of the most important representatives of realism in the 19th century and thus one of the involuntary ancestors of Socialist Realism, which was dictated by Stalin's policies in the 1930s. *Zaporizhian Kozaks Write a Letter to the Turkish Sultan* (1880–1891, St. Petersburg State Russian Museum) recreates a legendary scene from 1676. The Cossacks, who settled in the lower reaches of the Dnepr River, fought for their independence in the 17th century. Their state came into being after serf farmers fled from East Central and Eastern Europe at the turn of the 15th and 16th centuries and settled in what is now eastern Ukraine in the Kirovohrad, Dnipropetrovsk and Donetsk regions. At the beginning of the Ottoman-Russian War (1676–1681), Sultan Mehmed IV demanded the subjugation of the Zaporizhian Cossacks with a letter to their leader. Legend has it that the Cossacks also responded to this order with a letter, which was unusual for their customs. The letter is said to have consisted mainly of wild insults and culminated in a demand that the Sultan shall kiss their asses. (Friedman 1978: 29)

During Repin's time, the Cossacks were a role model for freedom and were admired by many, for example the Ukrainian-born author Nikolai Gogol, who recorded the stories about the Cossacks in his famous work *Taras Bulba* (1835). These have been adapted as operas and plays since the 19th century and as films several times since 1909. Although the legend about the letter and its exact wording remains controversial, it has had a strong impact on Ukrainian history. Many narratives in Ukraine are fed by the disobedience and love of freedom of the Zaporizhian Cossacks and the meme about the Russian warship can also be linked to this.

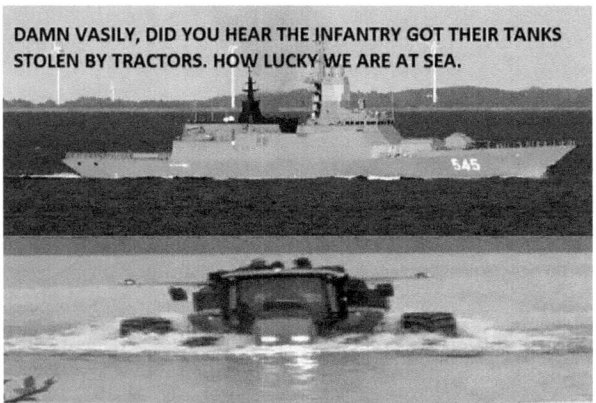

Figure 8: Ukrainian tractor is also dangerous in the water

The information about what happened to the border guards was contradictory until the end of March 2022, but with a meme or flash mob, the original information quickly loses its meaning. If we take a closer look at the iconography of the stamp, it reminds us of Chinese artist Ai Weiwei's famous series *Study of Perspective* (1995-2010), which he first started in Tiananmen Square in Beijing. His raised hand with the extended middle finger points towards the Gate of Heavenly Peace. This series was interpreted politically as a gesture of rebellion against the powerful and as an ironic gesture by a non-conformist artist, as the title refers to the process of adjusting the focus. To do this, people usually stick out their thumb to estimate the distances on cameras that do not have an automatic device for determining focus. In Ai Weiwei's photos, the gesture refers to a rebellious distancing of the photographing artist from the symbol of power depicted.

Tractors and Ukrainian Meme Force

The reference to the Cossacks and thus to the original and rural can be seen in another viral symbol, which goes back to a video that was posted on YouTube and Twitter from February 27, 2022. It showed a tractor pulling a Russian tank while someone runs behind trying to catch up. ("Russia-Ukraine War/Ukrainian Farmer 'Steals' Huge Russian Tank" 2022) "Ukrainian farmer steals Russian tank from right under the nose of the Russians who occupied it", says the subtitle.

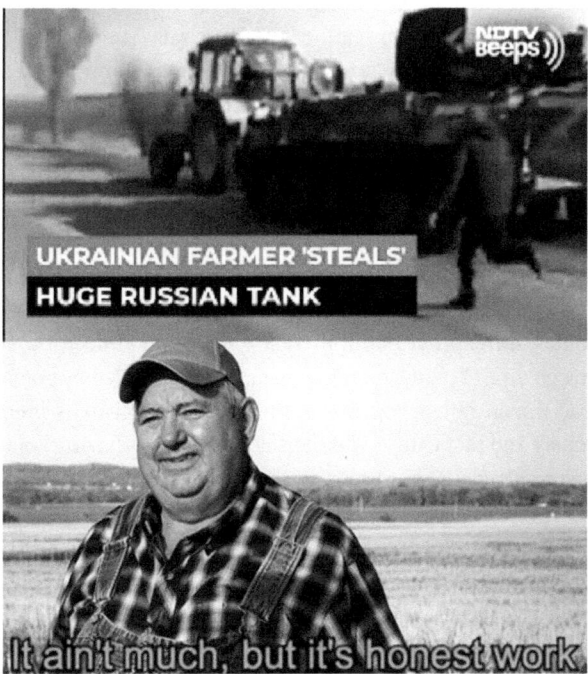

Figure 9: Memes on tractors and their drivers

Another video showed a Ukrainian from Berdiansk in the Zaporizhzhia region picking up a landmine with his hands and carrying it away from the road. (cf. Suciu 2022) Although established media felt compelled to determine the factuality of these contributions, they were completely uninterested in the memefication processes of such recordings. In the next step, they were converted into merchandising products, posters, and resistance advertising. In the meme universe, the question of "real or fake" is of secondary importance; virality decides. According to the David-Goliath formula, another topos was activated for the Ukrainian population with tractor drivers and farmers: that of the simple people from the countryside who, with wit and ambush, make the supposedly better-equipped invaders look stupidly.

Memes have become an important tool of warfare for Ukraine. In February 2022, an anonymous Twitter account "Ukrainian Memes Forces" (UKM) was set up. Here the pro-Ukrainian content is inserted into memes that have already gone viral or new memes that were created in Ukraine are posted. The content can be roughly divided into four categories: heroism and resilience of the Ukrainian population, mockery of Russian troops or Vladimir Putin, criticism of the lack of support from the West and the exposure of Russian fake news. (cf. Langschwager 2022) These themes emerged, among other things, from various Russian narratives that served as justification for the attack. The Russian government, in its usual twist of facts, stuck to the narrative of "denazification" of Ukraine, to which the Ukrainian side responded by calling the Russians "russists" (as equivalent of fascists). These narratives found a suitable transport medium in memes and, like war images, became an effective instrument of information warfare.

The evolutionary biologist Richard Dawkins introduced the term meme in his 1976 book *The Selfish Gene*. In it he defines it as a cultural analogue of the gene in biological evolution and coined the idea that memes develop and spread naturally and that users merely ensure their transport. There are two ways when it comes to spreading memes: on the one hand, there are people who are not aware that by spreading images or memes they are becoming part of a chain or are not interested in it. On the other hand, there are those who make it their mission to make a meme go viral. Finally, memes can also be controlled by bots. While funny memes spread uncontrollably online, opinions and criticism as memes are often used consciously and tactically – whether by creating or sharing content.

Memes not only unite, but they can also be exclusionary if they are based on a certain knowledge that is not accessible to every person. The Ukrainian memes use this tactic to win their "war of perception". Kohout sees Ukrainian memes as a striking recitation of Western pop culture. As an example a meme showing Bart Simpson writing "Russia is a Terrorist State" on the school blackboard, or a meme of Fred from the cartoon series *Scooby-Doo*, which reveals the villain Putin. (cf. Kohout 2022)

Bradley Wiggins looked qualitatively at the messages of the memes and found out that pro-Russian memes primarily used fecal language to label Ukraine and the European Union. In addition, pro-European and Western politicians were portrayed as zombies, Nazis and vampires. Ukrainian memes, of course, attacked the Putin cult and the idea of United Russia. As early as 2016, the frequent use of Western pop phenomena became noticeable, such as the protagonists from the films *Matrix* and *Austin Powers*. (cf. Wiggins 2016) The Ukrainian turn to Western popular culture has other reasons besides the signal of belonging: the use of universal Western imagery and English captions guarantee Ukraine more visibility in the West and also a more general understanding of the memes because they do not need to be explained.

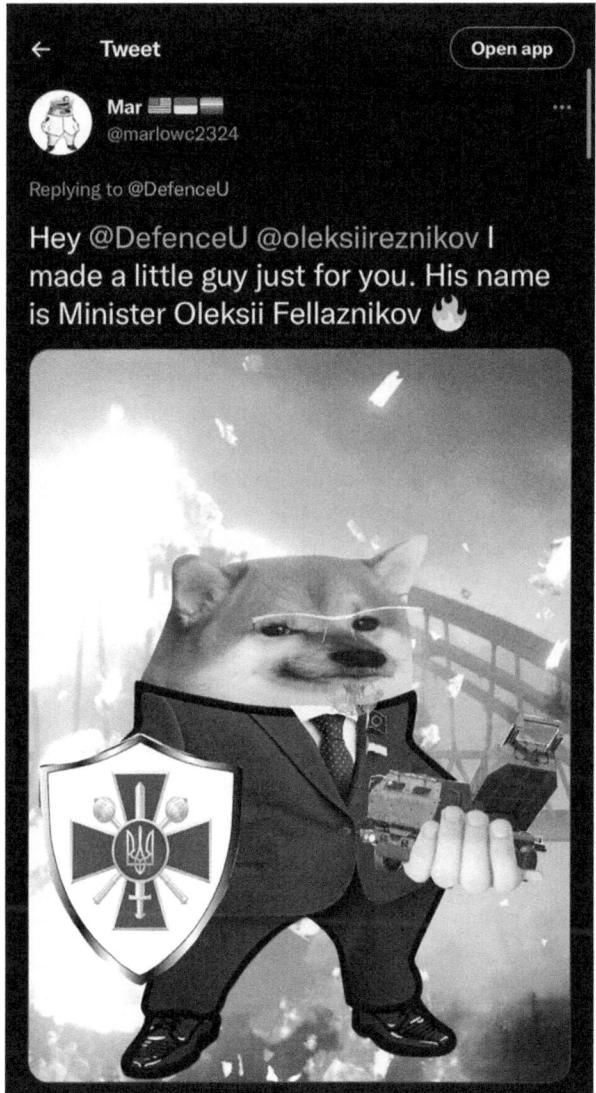

Figure 10: NAFO avatar of Ukrainian Defense Minister Oleksii Reznikov

There are also genuinely Ukrainian motifs. Saint Javelin, a Madonna with a US anti-tank guided missile Javelin in her hands, was made the patron saint of the Ukrainian army in 2022. (Straub 2023) This image was developed by Christian Borys before the Russian invasion began, but only later went viral. (Sommavilla 2022) There are also many memes in Ukrainian (visual) language, but they were hardly received in the West because there is a lack of background information. For example, geese can now be interpreted as "biological weapons", and the word "deadline" means a column of Russian military equipment. In addition, until a few years ago, Ukraine, like all countries of the former Soviet bloc, was heavily enriched with Soviet images and had to bring its own images, symbols and texts back to the forefront.

NAFO

The North Atlantic Fella Organization (NAFO, with a reference to NATO), which has been actively combating Russian propaganda and misinformation since May 2022, has become very well known. Their recognizable feature is the Japanese Shiba Inu dog, which has been used to transform propaganda material into memes and which is also used as the fellas' profile picture. (cf. Althaus 2022) Benjamin Tallis (security expert at the German Council on Foreign Relations) praised NAFO for its creative and funny strategy in the fight against Russian disinformation and Oleksii Reznikov (Ukrainian Defense Minister) thanked the group. (cf. ibid) This movement attracted thousands of people who did not know how to get involved against the war. NAFO's attacks are specifically directed against Russian propaganda and trolls. Their humor is infectious, like the warship meme. Their actions have certainly been crowned with success, such as the mockery of the Russian UN ambassador Mikhail Ulyanov:

The fellas had busted the man for his propaganda tweets in poor English. Ulyanov responded with "You pronounced this nonsense, not me," which subsequently became a widely used quote among the fellas. "The action even led to Ulyanov not posting anything on Twitter for a week," says Benjamin Tallis. (ibid.)

The NAFO runs a fan shop with sweatshirts, stickers, tickets for a fictional beach party in Crimea and collects donations for the Georgian Legion (a Georgian army in Ukraine) and the "Saint Javelin" project. All NAFO donors receive a Shiba Inu avatar. Ukrainian Defense Minister Reznikov also received a personalized fella after thanking NAFO on his Twitter account. Later he posted a post of himself wearing a fellas t-shirt. (Sommavilla 2022) Their reference to NATO goes beyond the name, so there is #NAFOarticle5, based on a similar article of the NATO Treaty, which offers each individual fella protection from the other fellas in an attack situation. (Althaus 2022) They do a lot in their field of social media and raise the question: Are their actions in their field more effective than those of NATO?

The fellas' recipe for success is the logic of viral memes: They use a fluffy pet that has been known for years as the "Doge" meme and has also existed in the form of a cryptocurrency since 2013 as "Dogecoin". So they enrolled in an already successful system to get more attention. With the cute meme they demonstrate the already clumsy Russian propaganda and let the carnivalesque laughter of Mikhail Bakhtin echo through the Internet. Another factor is their mass; the many postings allow them to oppose the Russian propaganda machine. Because everyone can participate, a whole new sense of community is created that drives this movement. With their actions, the fellas refute the idea that there is only "click activism" on social media, i.e. reactions to postings that have no use in the real world, says Olga Boichak, lecturer in digital cultures at the University of Sydney. She sees NAFO as an example of "participatory warfare", which "can directly or indirectly influence the outcome of a military conflict on site". (Oertli 2022) It is an exciting question as to what the future will do to NAFO, how it will develop as a group and deal with external influences.

Russian Memes

The pro-Ukrainian memes are clearly in the lead. Not only because there are more successful, funny memes among pro-Ukrainian memes. They also have more platforms and distribution channels. And this is a very important aspect of meme culture and the internet in general. Many platforms have blocked all uploads for all users from Russia. This makes it more difficult for them to distribute their content. Russia itself blocks social networks for its citizens. Of course, the blocks can be circumvented, but the number of active users is declining. (Straub 2023)

Which memes are produced in Russia and what is their virality? The Z-propaganda, which appeared on all social media channels since February 2022 and spread perfidious news, is still available online under hashtags such as #russianlivesmatter or, for short, #rlm. There is a lot of Z-content on TikTok, such as memes and videos, which are mostly in Russian and set to accompanying pop music, or even patriotic folk songs and hymns. It proceeds according to the tried and tested principle of patriotism, nationalism, exaggeration of enemy images and trivialization or justification of the "special military operation". These memes are hardly received in Europe, but just as often as the Ukrainian ones, they make use of the existing meme arsenal, such as the collage, which has been used as a meme for years to show various comparisons. Reactions to different wars are shown here: While the family in front of the screen says "We must stop them" when the USA attacks Libya, Pakistan or Syria, there is suddenly a protest when Russia attacks Ukraine. This portrays the West as hypocritical.

The Russian memes against the war are also in Russian language and often use finished material from *The Simpsons*, the popular series *Friends*, but also many photographs of Russian politicians. Outside Russia, they can only be understood by Russian-speaking audiences and often refer to genuinely Russian cartoon characters and stars. The demarcation mechanism is more effective here than with Ukrainian-language memes – at the same time, the Russian-speaking population, who often do not speak any other language, is intended to be addressed and activated.

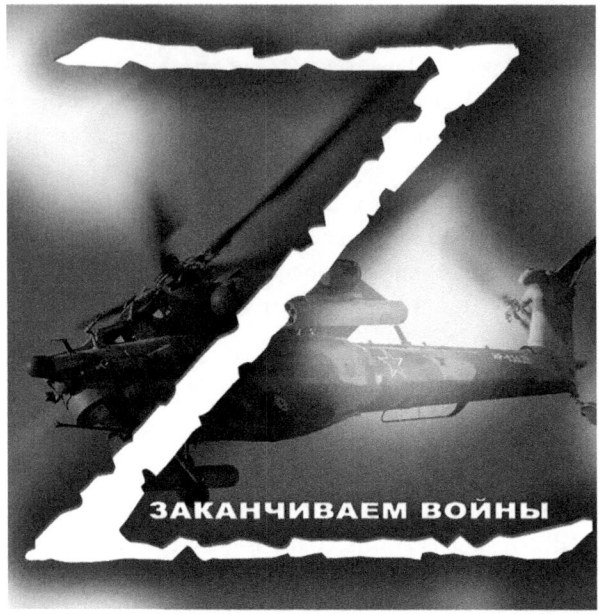

Figure 11: *"Let's end wars", meme from the Ministry of Defense of the Russian Federation from March 5, 2022. (Here the cyrillic letter "з" is stressed with the letter "z", but not substituted yet. This way of writing spread in Russia in 2022. In the region Kusbas officials changed the writing to "KuZbas" to show their alliance with Russian war propaganda. cf. Iljina 2022.)*

A meme or rather a comic shows Leopold the cat from the Soviet-Russian animated series of the same title. Since its creation in the 1970s, the cat has worn a yellow shirt and blue pants – coincidentally the colors of the Ukrainian flag. To demonstrate the absurdity of the current measures, the cat in the meme is beaten up by three police officers. Because expressions of solidarity with Ukraine are seen in Russia as "discrediting the Russian army". (cf. Kohout 2022) You must know that the cat Leopold has always been pestered by two mice and his recurring call, which he repeats in every episode – "Guys, let's live peacefully" – always leads to the same result: he is attacked and beaten by the mice. This meme was created by Belarusian artist CHILIK (@chiliktol), who posts anti-war comics on his Instagram account.

When the war goes viral 153

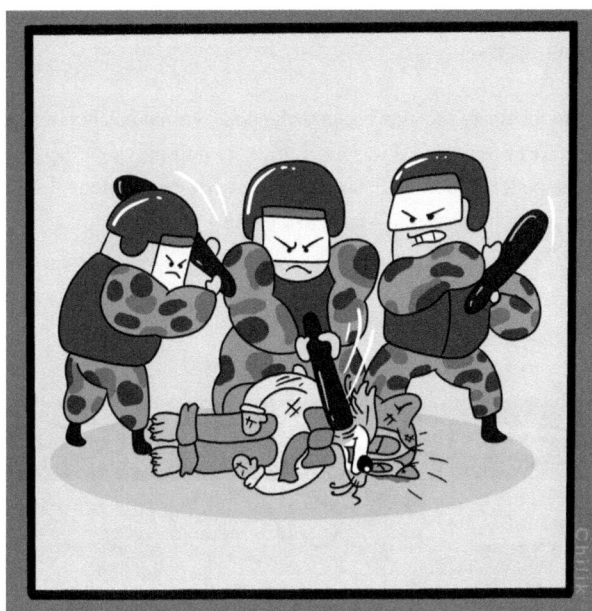

Figure 12: CHILIK: Cat Leopold and the police

Kohout refers, among other things, to the meme "Хуй войне!", which is still a reference from the early 2000s in Europe. The slogan translates as "War fuck you" or "Fuck War" and was worn on t-shirts by the Russian girl band t.A.T.u. during an appearance on *The Tonight Show* in 2003. Even though they had been prohibited to comment on the war in Iraq they decided to wear their protest live on television. The Bush administration banned protests against the war during the Grammy Awards, but they happened anyway, because as a popular figure in the US you could be sure that nothing would happen if you violated it. Unlike in Russia, which is why there are fewer anti-war memes from there. Since 2015, it has been forbidden to produce memes of celebrities and the spread of so-called denigration of the armed forces and the fatherland is also a punishable offense. (cf. Kohout 2022)

The slogan "Хуй"-Something was not invented by t.A.T.u. and has always existed in one form or another in the Russian-speaking protest culture. For example, in actions by Moscow artists in the 1990s that are worth remembering in this context. The non-conformist artist group E.T.I. laid out the three-letter word on the Red Square, using the bodies of group members and people they persuaded on the street to join in as letters. The idea behind it was to finally start a direct conflict with the government and to criticize the economic and political situation. It was also directed against the so-called "morality law", which meant imprisonment for up to two weeks if someone used vulgar language in public places (it came into force on April 15, 1991). Anatoly Osmolovsky, one of the group-members, remembers:

We only lay there for thirty seconds because the police immediately ran to us and started pulling our hair; They then said it was even funny to pull the "x **" up by his hair. They dragged us to the police station and started asking what we had done. I said that we have shown different geometric forms. (Osmolowskij 2016.)

Mogilisazia

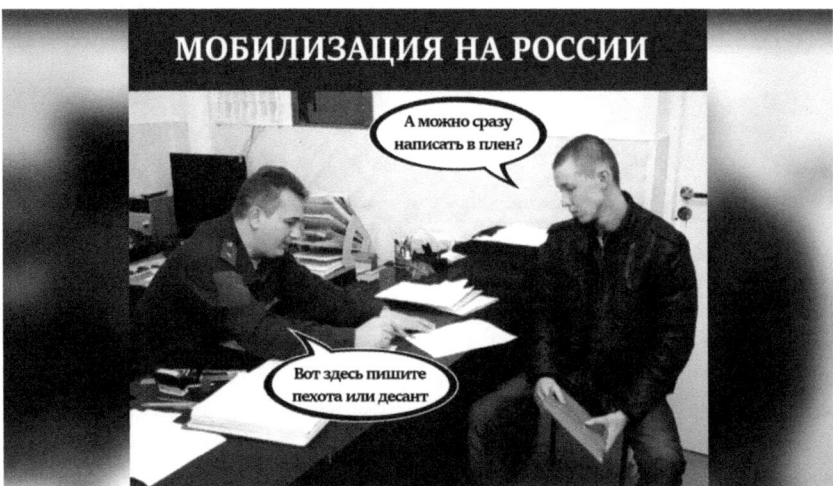

Figure 13: Here you can write whether ground troops or air troops. – Can I write captivity immediately?

Many memes have been produced since the September 2022 against Russian mobilization. The Russian word "Mobilisazia" became "Mogilisazia" [for mogila: tomb], alluding to the senselessness of this war on the one hand and the impending death on the other – both factors are a criticism of the Russian leadership and the military. These and many other memes are barely visible in Germany, while Ukrainian memes, videos, tweets and images of all kinds are omnipresent. They should rouse and mobilize against the war. Russian propaganda is aimed at the Russian-speaking world and excludes everyone else, so that at best they stay out of it. The Russian-speaking opposition is in a difficult position between these two sides – here we come back to the question: who is allowed to joke? Who can laugh?

This question is often discussed in German and English-speaking media. How relevant is humor in a situation of real human suffering? I think a lot depends on the position of the actors. Because it's one thing when Ukrainians post funny memes about the invasion of their country, and quite another when the posting people are safe and far away from the battlefields. Even if they do it as a gesture of solidarity with Ukraine. (Straub 2023.)

The lack of memes from Russia's opposition should also be understood as a gesture of solidarity with Ukraine. The news channel *Meduza* published first memes on October 4, 2022, to show the reactions to Elon Musk's statements about the war. One of the world's most privileged man tweeted a list of conditions under which he believed the two fighting states can make peace. Musk proposed "keeping Crimea, which was annexed in 2014, as part of the Russian Federation", holding a new vote in the occupied territories and recognizing Ukraine as a "neutral" state. The memes published by *Meduza* were under the following heading: "During the war we publish (almost) no memes. But now we are forced to make an exception. Elon Musk is in big trouble: he has decided to say something about the war. The Internet's reaction." ("Vo vremia voiny my (praktitscheski) ne publikuiem memy. No seitschas vynushdeny sdelat iskliutschenije. U Ilona Maska bolschije problemy: on reschil vyskazatsia po povodu voiny. Reakzia interneta" 2022.)

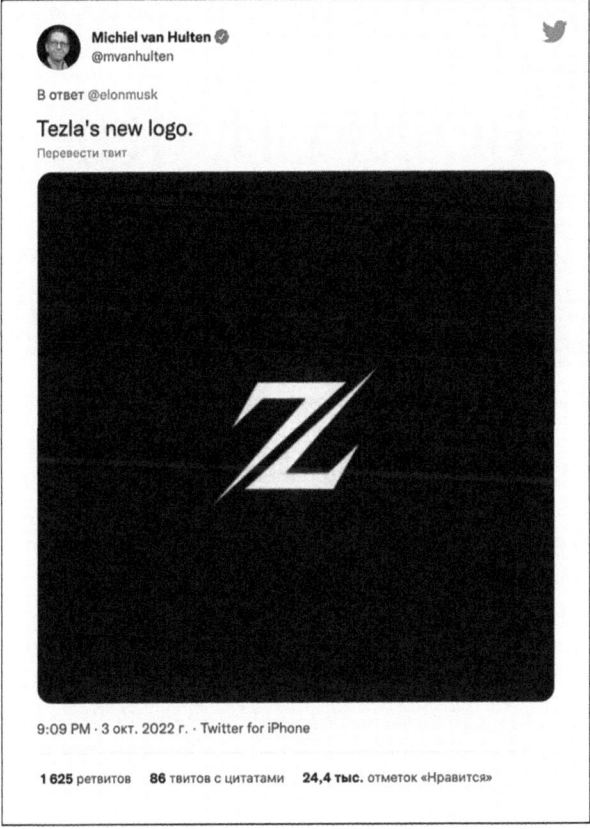

Figure 14: Meme on Tesla, Elon Musk and his ideas how to end the war

No to dried fish

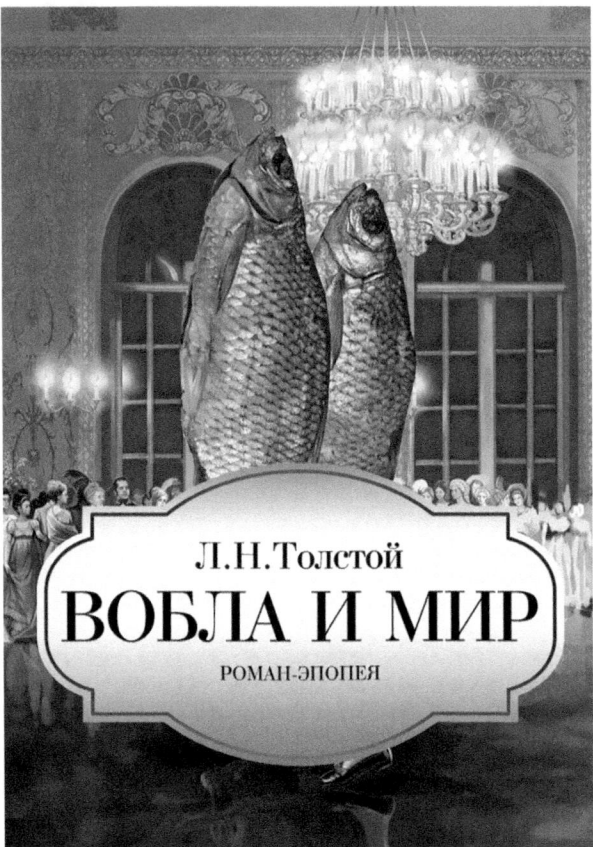

Figure 15: Leo Tolstoi "Vobla and Peace", altered book cover

The story of "Net Voble" proves that the Russian opposition still exists and proves a sense of humor. In Tyumen, on October 12, 2022, the court heard the case of 30-year-old Alisa Klimentova, who wrote "Net ****e" in chalk on the sidewalk in a city square, which usually stands for "Net Voine". (No to war). The defendant was able to prove that she meant "No to Vobla" [dried fish traditionally eaten with beer and vodka] and not "No to War" because she had a strong aversion to this type of fish. The case went viral on social networks, and the dried fish became an anti-war meme. Internet users replaced the word "war" with "vobla" in book, film and song titles. The famous Russian singer and songwriter Semyon Slepakov wrote a song about "Net Voble" with clear statements against the war in Ukraine and Ksenia Sobchak created a capsule collection with the slogan. In Russia, for a few weeks, the fish replaced the dove of peace as the symbol of an anti-war movement. Just a month after the first hearing, Klimentova's criminal case was brought back to court. ("Sud otmenil reschenije po delu 'Net voble'" 2022.) She was ultimately sentenced to a fine of 30.000 rubles – this sum was raised for her through crowdfunding. ("Tjumenzy za sutki sobrali 30 tysjatsch rublej, tschtoby pogasit schtraf figurantke 'dela o voble'" 2022)

These stories make it clear that many anti-war memes are being produced in Russia, whether from existing meme material or newly created, like the fish. These memes are not accessible to non-Russian-speaking people and the channels on which they are distributed are also more Russian, such as vKontakte, Odnoklassniki, Telegram, etc. The developments of the information war show how Ukraine and the rest of the world (see NAFO) has learned from Russia's formerly seemingly invincible propaganda, troll and hacking machinery. The future will certainly bring to light many currently hidden attempts to express oppositional opinions in Russia, as well as the Ukrainian-language memes with a clear nationalist character, which also exist but are not currently the center of attention.

Meanwhile, an exhibition of memes about the Ukraine War was held in the Berlin Story Bunker, where the topic of the German capital during the Second World War, the Cold War and the period afterward are usually the main exhibition topics. The show was curated jointly by the museum directors Wieland Giebel and Enno Lenze. They showed the NAFO, Zelenskyy as Captain America and claim that this is the world's first meme exhibition. (Rushton 2022) Putin, the Meme, as he was shared and exhibited in 2015 is already forgotten.

References

Althaus, Peter. "Russische Desinformation. NAFO Fellas: Bewegung von Meme-Hunden zerstört die Putin-Propaganda", Berliner Kurier, September 6, 2022.
Balforth, Tom/Trevelyan, Mark/Jones, Gareth (2022): "'Russian warship, go fuck yourself': Kyiv to honour troops killed on island", Reuters; February 25, 2022.
Bidder, Benjamin. Generation Putin. Hamburg: DVA, 2016.
"Da legt der Knirps den großen Putin aufs Kreuz", Der Spiegel, February 25, 2023. Accessed 3 December 2024: https://www.spiegel.de/kultur/banksy-briefmarke-in-der-ukraine-die-kleine-ukraine-legt-das-grosse-russland-aufs-kreuz-a-12a22dc6-1c62-4937-8556-cf66366f3663.
Eilenberger, Wolfram. Wolfram Eilenberger im Gespräch mit Charlotte Klonk und Wolfgang Ullrich: "Bilderkrieg und Bilderkrise", Deutschlandfunk Kultur, March 06, 2022. Accessed 3 December 2024: https://www.deutschlandfunkkultur.de/bilderkrieg-ukraine-100.html.
Friedman, Victor A. "The Zaporozhian Letter to the Turkish Sultan: Historical Commentary and Linguistic Analysis", in: Slavica Hierosolymitana 2, 1978, pp. 25-38.
von Gehlen Dirk. Memes. Berlin: Wagenbach, 2020.

Iljina; Anastasia. "Vlasti Kemerovskoi oblasti stali nasyvat region KuZbassom v snak poddershki spezoperazii na Ukraine", NGS.ru, March 02, 2022. Accessed 3 December 2024: https://ngs.ru/text/gorod/2022/03/02/70480919/.

Kohout, Annekathrin. "Katzen, Pop und Kriegsschiffe", Zeit Online, June 04, 2022. Accessed 3 December 2024: https://www.zeit.de/kultur/2022-06/memes-russland-ukraine-krieg-social-media?wt_zmc=sm.int.zonaudev.twitter.ref.zeitde.redpost.link.x&utm_medium=sm&utm_source=twitter_zonaudev_int&utm_campaign=ref&utm_content=zeitde_redpost_link_x.

Langschwager, Marit. "Zwischen Superhelden und Schurken: Die Macht der Memes im Ukraine-Krieg", Neue Zürcher Zeitung, June 25, 2022. Accessed 3 December 2024: https://www.nzz.ch/technologie/humor-spott-und-zynismus-die-macht-der-memes-im-ukraine-krieg-ld.1690064.

Masalzeva, Maria. "Vystavka Memov s Putinym proidet v Moskve nakanune jego dnja roshdenija", Afisha Daily, October 10, 2017. Accessed 3 December 2024: https://daily.afisha.ru/news/11562-vystavka-memov-s-putinym-proydet-v-moskve-nakanune-ego-dnya-rozhdeniya/.

Medvedev, Sergei. "Krieg im Namen des Sieges von 1945", dekoder, April 27, 2022 Accessed 3 December 2024: https://www.dekoder.org/de/article/krieg-ukraine-9-mai-tag-sieges.

Medvedev, Sergei. Park Krymskogo Perioda. Chroniki tret'ego sroka, Moscow: Individuum, 2017.

Mishchenko, Taras. "Ukrposhta announces new 'Russian Warship...DONE' military stamp", Mezha.Media, April 22, 2022. Accessed 3 December 2024: https://mezha.media/en/2022/04/22/russian-warship-done-military-stamp/.

Neuhaus, Andreas. "Dogecoin. Spaßwährung legt 8000 Prozent zu – Experten ratlos: 'Was sich da abspielt, ist komplett irrational'", Handelsblatt, April 20, 2021. Accessed 3 December 2024: https://www.handelsblatt.com/finanzen/maerkte/devisen-rohstoffe/dogecoin-spasswaehrung-legt-8000-prozent-zu-experten-ratlos-was-sich-da-abspielt-ist-komplett-irrational/27111594.html.

Oertli, Sandro. "Die Nafo – eine 'Meme-Armee' in Kampf gegen russische Propaganda", Tagesanzeiger, September 01, 2022. Accessed 3 December 2024: https://www.tagesanzeiger.ch/die-nafo-eine-meme-armee-im-kampf-gegen-russische-propaganda-286788953636.

Osmolowskij, Anatolij. "Akzii 'E.T.I. – tekst' – 25 let!", Artgid, April 18, 2016. Accessed 3 December 2024: https://artguide.com/posts/1019.

Petrenko, Roman. "'Rosijskij poizd, idi na khui!': usi zaliznichni vuzli z Rosieiu znisheni", Ukrainska Pravda, February 26, 2022. Accessed 3 December 2024: https://www.pravda.com.ua/news/2022/02/26/7326208/.

Potschepzov, Georgii G. Psichologitscheskije voiny, Moscow: Refl-buk, 2000.

Rushton, Elizabeth. "Diese Ausstellung zeigt, wie die Ukraine russische Aggression mit Memes bekämpft", Berliner Zeitung, November 19, 2022. Accessed 3 December 2024: https://www.berliner-zeitung.de/kultur-vergnuegen/

diese-ausstellung-zeigt-wie-die-ukraine-den-krieg-mit-memes-gewinnen-will-li.288438.

"Russia-Ukraine War/Ukrainian Farmer 'Steals' Huge Russian Tank", NDTV, March 01, 2022 Accessed 3 December 2024: https://www.youtube.com/watch?v=yIB7QPsNA_w.

"Russland verbietet Memes. Spaßbremse des Internets", Stern, April 15, 2015. Accessed 3 December 2024: https://www.stern.de/russland-verbietet-memes--die-spassbremse-des-internets-6963712.html.

Sauer, Pjotr: "Ukraine gives medal to soldier who told Russian officer to 'go fuck yourself'", The Guardian, March 29, 2022. Accessed 3 December 2024: https://www.theguardian.com/world/2022/mar/29/ukrainian-soldier-russian-warship-medal-snake-island.

Schomowa, Swetlana. "Ein Putin – Viele Memes", dekoder, May 05, 2020. Accessed 3 December 2024: https://www.dekoder.org/de/article/putin-memes-propaganda-populaerkultur.

Semotiuk, Orest. "Ukraine: Humor als Waffe im Krieg", ZOiS Spotlight 2/2023, January 25, 2023. Accessed 3 December 2024: https://www.zois-berlin.de/publikationen/zois-spotlight/ukraine-humor-als-waffe-im-krieg.

Shuster, Simon. Vor den Augen der Welt. Wolodymyr Selenskyj und der Krieg in der Ukraine, Munich: Goldmann, 2024.

Sommavilla, Fabian. "Memes für die Propagandaschlacht: Die unterschätzte Waffe im Arsenal der Ukraine", Der Standard October 21, 2022. Accessed 3 December 2024: https://www.derstandard.de/story/2000140121857/memes-fuer-die-propagandaschlacht-die-unterschaetzte-waffe-im-arsenal-der.

Straub, Verena. "Die pro-ukrainischen Memes liegen eindeutig in Führung", Süddeutsche Zeitung, January 27, 2023. Accessed 3 December 2024: https://www.jetzt.de/ukraine-krieg/krieg-gegen-die-ukraine-welche-rolle-memes-spielen.

Suciu, Peter. "Real or Fake: Video Of a Farmer Stealing Russian Tank, Landmine Removed With Bare Hands, And Comparisons Of Putin To Hitler Trading", Forbes, March 02, 2022. Accessed 3 December 2024: https://www.forbes.com/sites/petersuciu/2022/03/02/real-or-fake-video-of-farmer-stealing-russian-tank-landmine-removed-with-bare-hands-and-comparisons-of-putin-to-hitler-trending/?sh=15a875cb43ec.

"Sud otmenil reschenije po delu 'Net voble'", Lenta.ru, November 21, 2022. Accessed 3 December 2024: https://lenta.ru/news/2022/11/21/net/.

"Tjumenzy za sutki sobrali 30 tysjatsch rublej, tschtoby pogasit schtraf figurantke 'dela o voble'", fontanka.ru, December 21, 2022. Accessed 3 December 2024: https://www.fontanka.ru/2022/12/21/71918045.

Ullrich, Wolfgang. "Zum Krieg und den Bildern Kulturwissenschaftler Wolfgang Ullrich: 'Selenskij schreibt Mediengeschichte'", Deutschlandfunk, March 03, 2022. Accessed 3 December 2024: https://www.deutschlandfunk.de/koennen-bilder-einen-krieg-beenden-kulturwissenschaftler-wolfgang-ullrich-dlf-54046f5f-100.html.

Visontay, Elias (2022): "Ukraine soldiers told Russians to 'go fuck yourself' before Black Sea island death", The Guardian, February 25, 2022. Accessed 3 December 2024: https://web.archive.org/web/20220225102114/https://www.theguardian.com/world/2022/feb/25/ukraine-soldiers-told-russians-to-go-fuck-yourself-before-black-sea-island-death.

"Vo vremia voiny my (praktitscheski) ne publikuiem memy. No sejtschas vynushdeny sdelat iskljutschenije. U Ilona Maska bolschije problemy: on reschil vyskazatsia po povodu voiny. Reakzia interneta", Meduza, October 04, 2022. Accessed 3 December 2024: https://meduza.io/feature/2022/10/04/vo-vremya-voyny-my-prakticheski-ne-publikuem-memy-no-seychas-vynuzhdeny-sdelat-isklyuchenie.

Vollmer, Jan. "Wladimir Putin ist der neue Chuck Norris", Welt, March 25, 2014. Accessed 3 December 2024: https://www.welt.de/vermischtes/article126167791/Wladimir-Putin-ist-der-neue-Chuck-Norris.html.

Wiggins, Bradley E. "Crimea River: Directionality in Memes from the Russia-Ukraine Conflict", in: International Journal of Communication 10 (2016), pp. 1-34.

Zygar, Mikhail. Krieg und Sühne. Der lange Kampf der Ukraine gegen die russische Unterdrückung, Berlin: Aufbau, 2023.

Biographical Notes

Tatsiana Astrouskaya (PhD) is researcher at the Herder Institute for Historical Research on East Central Europe and lecturer of Digital History at the University of Giessen. Her research focuses on the history of cultural and political opposition in Eastern Europe, policies and practices of digitalization, and post-Soviet memory politics. She is the author of award-winning Cultural Dissent in Soviet Belarus (1968-1988) (Harrassowitz 2019, Viasna 2022, Novoe literaturnoe obozrenie 2024). Currently, she is working on her second book project on Jewish emigration and the challenges of socialist modernity in the Soviet peripheries.

Max Kramer (PhD) is senior researcher at the Department of Anthropology, Freie Universität Berlin. He currently investigates the digital politics of Indian Muslims within the project *The Populism of the Precarious* funded by Volkswagen Foundation. He is the author of *Mobilität und Zeugenschaft* (Transkript 2019), a book that investigates independent documentary film practices which deal with the Kashmir Conflict. Max translates Hindi literature, makes documentary films and is co-editor in chief of the journal *Dastavezi: The Audio-Visual South Asia*.

Maksim Markelov is a PhD candidate at the University of Manchester carrying out research under the supervision of Professors Stephen Hutchings and Vera Tolz. Maksim's PhD project entitled *Transforming Meaning: Russian Trolls in Social Media's Changing Linguistic Landscape* uses a combination of quantitative and qualitative methods and aims to provide the first substantive analysis of how the linguistic practices of online actors identified as Russian state-sponsored trolls change with time and context.

Lesia Kulchynska (PhD) is a curator and visual studies researcher currently fellow at the Netherland Institute of Advanced Studies where she works on the visuality of violence during the Russo-Ukrainian war. She has taught at the National University of Kyiv-Mohyla Academy (Kyiv), Kyiv Academy of Media Arts", and John Cabot University (Rome). She also worked as a researcher at the Pinchuk Art Center and curator at the Visual Culture Research Center in Kyiv. She did a postdoc at Bibliotheca Hertziana. Max Planck Institute of Art History (Rome) 2022-2024. Recent publications: *Meaning Production in Cinema: Genre Mechanisms* (2017), editor of *The Right to the Truth: Conversations on Art and Feminism* (2019) and *Joseph Beuys. Everyone is an artist* (2020).

Yana Lysenko is a researcher at the Research Centre for East European Studies, working on the research project "Comparing Protest Actions in Soviet and Post-Soviet Spaces, Data Reuse - Ukraine," funded by the Volkswagen Foundation. During the current winter semester 2024/25, she is teaching as a lecturer at the University of Bremen in the course "Introduction to Comparative Political Science." Her research expertise focuses on conflict studies, particularly Ukraine-Russia relations, de facto states, and the dynamics of identities and civil society in times of crisis. In her Master's thesis, entitled War in the Minds, she explored identity as a factor influencing the perception of war, using the case of the "Donetsk People's Republic."

Serhii Zasiekin (DSc) is researcher in the School of Slavonic and East European Studies (SSEES) at University College London, UK, and Professor in the Department of Applied Linguistics at Lesya Ukrainka Volyn National University, Ukraine. He received the 2024 Nevitt Sanford Award for Outstanding Professional Contribution to Political Psychology for the *Narratives of War* project. His research interests include the ethics of translation, translation universals, psycholinguistics of Bohdan Lepky and present-day Ukrainians' war narratives. He is currently Vice-President of the Ukrainian Association of Psycholinguists. He is also Editor-in-Chief of *East European Journal of Psycholinguistics*.

Elena Korowin (PhD) is teaching art history with a focus on contemporary art at the Braunschweig Academy of Fine Arts. She was awarded the ifa Research Prize for her dissertation *The Russian Boom. Art exhibitions as a means of diplomacy between the Soviet Union and the Federal Republic of Germany 1970-1990* at Karlsruhe University of Arts and Design in 2015. In addition to many curated exhibitions and active work as an art critic for various newspapers and magazines, her research focuses on art and politics, art autonomy, feminism and gender, and postcolonial studies.